travel

SKY
DANCE

ALSO BY JOHN D. BURNS
The Last Hillwalker
Bothy Tales

SKY
DANCE

*Fighting for the wild in the
Scottish Highlands*

JOHN D. BURNS

Vertebrate Publishing, Sheffield
www.v-publishing.co.uk

First published in 2019 by Vertebrate Publishing.

VERTEBRATE PUBLISHING
Omega Court, 352 Cemetery Road, Sheffield S11 8FT, United Kingdom.
www.v-publishing.co.uk

Cover illustration by Jane Beagley.

Edited by Pinnacle Editorial.
www.alexroddie.com/pinnacle-editorial

A CIP catalogue record for this book is available from the British Library.

ISBN 978-1-912560-26-4 (Paperback)
ISBN 978-1-912560-27-1 (Ebook)
ISBN 978-1-912560-28-8 (Audiobook)

10 9 8 7 6 5 4 3 2 1

Production by Vertebrate Publishing.
www.v-publishing.co.uk

Vertebrate Publishing is committed to printing on paper from sustainable sources.

Printed and bound in Great Britain by Clays Ltd, Elcograf S.p.A.

CONTENTS

~

For the land and its forgotten creatures.

They hang the man and flog the woman
That steal the goose from off the common,
But let the greater villain loose
That steals the common from the goose.
The law demands that we atone
When we take things we do not own,
But leaves the lords and ladies fine
Who take things that are yours and mine.

—English nursery rhyme, circa 1764

~

NOTE ON THE TEXT

The Isle of Morvan is a fictional location, albeit inspired by the geography, ecology and culture of the Scottish Highlands and Islands. The Muir and Purdey estates are also fictional, as are all the characters in this story with the exception of public figures such as First Minister Nicola Sturgeon. But while this might be a work of fiction, that doesn't mean it isn't true.

PROLOGUE

In the darkness the Land Rover lurched violently around a bend in the narrow track, and the young man in the passenger seat grabbed the door handle to brace himself. His hand was sweating and he noticed that his fingers trembled. He had expected to be calm and determined, but instead his stomach churned at the thought of what he was about to do.

He turned to the older, grey-bearded man at the wheel, who stared hard through the windscreen into the blackness of the night, and fought to push back the panic. What if this whole thing ended in disaster? What if they were caught? There were over twenty years between himself and the driver – the older man had built a career, people respected him, and all that might be lost in the next few hours.

The silence between them over the last few hours had become solid and the young man struggled to break it. 'We can still turn back if you don't want to go through with it.'

As soon as he said the words, he knew they were a lie. They had gone too far to turn back now. The older man did not answer, did not even seem to be listening.

The young man spoke again. 'I don't think I could do prison.' That was the truth.

When at last the older man spoke he was calm and determined, his accent revealing his east Scotland roots. 'All my life I've done richt by the rules, waited, been patient.' Now he turned, passion in his eyes. 'I want to live to see it, yer ken, and unless we do this I won't, maybe you won't. They'll stop us.'

He wrung his hands on the wheel, as if trying to mould the rage in his head into words. 'I want to bring a bit of wildness back into the hills. I'm not going to wait. Let them try and stop me.'

They drove on into the night.

At last the Land Rover slowed at the gated entrance to a track and came to a stop, the wire mesh cage in its trailer rattling. The driver leaned forward and switched off the headlights. Night swallowed them and for a moment the only sound was the throbbing of the old diesel engine. When the driver turned the key, the hum of the engine ceased. Now silence flooded out of the blackness.

The younger of the two men, still in his thirties, his body lean and taut, his hair spiked and ginger, scratched his beard and stared into the night. This whole thing had been his idea. It had taken him weeks to convince the older man to do this, yet now they were at the gates, now the moment had come, doubts wracked him. The older man lived by rules, never stepped out of line, always argued for caution. But the young man had been persuasive. Slowly the older man had changed. Now it was him who was full of fire; now it was the younger man who felt the hand of caution on his shoulder.

'Well?' he said out loud, but the question was for himself. 'Well, what now?'

The older man did not reply. The pair sat for a moment and, as their eyes adjusted to the darkness, the outline of the metal farm gate took shape before them.

'You ken whit to do,' the older man replied, cutting off the question.

'OK, OK.'

No going back. The younger man's heart pounded against his ribcage. Neither man moved.

There is a stillness in some moments when time stands before you, its surface smooth and still. You can stand and see your life reflected in that surface with a pebble in your hand, knowing that the instant you throw it, time will ripple away and you will be changed forever.

The older man reached under the dashboard for a head torch. He turned to the younger man. 'Go on, then.'

The young man took the torch, hands shaking. An instant later dazzling light filled the Land Rover and cast the shadows of the two men out on to the surface of the track.

The older man all but jumped out of the driving seat. 'Christ, for God's sake!'

The younger man was frantically pressing the switch. 'I had it on moonlight.' He jabbed at it again and a vivid stroboscopic light caught the occupants of the four-wheel drive in frozen moments like dancers in a bizarre ballet.

The older man pressed his face into his hands. 'Why don't yer phone the press while yer at it, man?'

To the relief of both the light faded to a subdued glow and they were left breathing heavily, staring out into the darkness for a sign that someone was coming.

The young man began to see indeterminate shapes as his eyes struggled to adjust to the faint light.

'Someone's out there, look!' the younger man said in a hoarse whisper.

The older man turned and followed his gaze. He held his breath, reached out and took hold of the ignition key. They both froze. Moments later a sheep trotted past the Land Rover, pausing only to sniff the scent of the men before vanishing once more.

'Christ.' The young man fell back into his seat, sighing with relief and tugging nervously at his beard. 'I thought—'

The driver shook his head in frustration. 'Get on with it.'

The young man grunted, bent over and produced a pair of heavy bolt cutters from beneath the passenger seat. In the dim glow of the head torch the two men exchanged glances. They both knew they were about to hurl the pebble.

Out of the confines of the Land Rover the air was cold and the young man's breath misted in the beam of his head lamp as he fixed the teeth of the bolt cutter to the chrome of the lock's hasp. An image flashed through his mind of a fox with its jaws around the neck of a swan. He shrugged, shook the vision from his mind and squeezed the bolt cutter's handles together. For a moment the lock resisted; his arms and shoulders trembled with the force. There was a sudden crack and the lock hit the gravel of the track, defeated. The young man glanced around, fearing the sound might have attracted attention, but only the blackness stared back at him. The galvanised steel of the gate shrieked as it swung open, and somewhere, not too far away, a dog barked. The young man felt a tremor creep up his spine at the sound.

The Land Rover's engine burst into life and the older man leaned out of the window. 'Come on, for fuck's sake.'

The young man leapt into the rattling vehicle and they set off along the track using only the meagre torch beam to guide them. As they passed through the gate the pebble arced through the air and began to spin as it made its descent. It hit the surface and time rippled out, its stillness broken forever.

In the pale beam of the head torch, trees lurched in and out of view and fence posts appeared with alarming speed.

'Turn left,' the young man urged.

The older man struggled with the gear lever. 'I cannae see.'

'I know where we are – turn left.'

The Land Rover swung round, a tree appeared and the driver turned the wheel and braked. He was quick but the tree was quicker and managed to catch a wing. The car shuddered; glass tinkled on to the driveway.

'What? Oh Christ!' The older man rammed the gears into reverse.

'Bugger, I thought we were further on.'

'You couldnae find your way doon a garden path. Point the bloody torch oot there. I can't drive in the damn dark.'

A gate appeared on the left in the glow of the head torch, and the young man exclaimed in triumph. 'See, see, there it is. I do know where we are going.'

'You ken where we were going when you told me to turn left and we hit that bloody tree?'

'There's no need to be all sarcastic. We can turn around now if you like.'

Out in the blackness a pair of eyes caught in the glow shone back at them. Then another pair appeared, quickly followed by a third and then a fourth.

The young man whispered, 'Deer, that's all they are – deer.'

The older man hissed as if he were a tyre and someone had just let all his air out. 'Sure of that, are yer? Mind where we are. They got bears here.' Then he added, as an afterthought, 'And tigers too.'

The risks of breaking in to an animal park were not lost on the younger man. 'Just drive.'

The car swung away and the eyes followed. At last the torch beam highlighted the chain links of a high wire-mesh enclosure, white against the darkness. The two men struggled with the coupling of the trailer for a few seconds before freeing it from the Land Rover and turning the cage so that its rear door faced the gate of the enclosure. They pushed the trailer up the embankment that led to the enclosure door, tyres sinking in to the soft ground. Both men were panting with the effort by the time the door of the trailer met the mesh of the enclosure wall. The two doors clashed and the mesh shuddered with a metallic scream for a few seconds after the impact.

'Shit,' the older man gasped, sucking in air after the effort of pushing the trailer.

As if in answer, the primeval howl of a wolf filled the air around them. Another wolf then another followed. The howls reached down into the bodies of both men and tightened the walls of their stomachs. Darkness and wolves are brothers and both bring a chill to the hearts of human beings from a time our minds have forgotten but our bodies know well – a time not so long ago, when to stray from the firelight was to walk with death, a time when the cry of the wolf rippled through men's souls.

The young man pushed his sleeve back and peered at his watch. 'Three fifteen. We've still time.'

The older man, his beard dewed with sweat in the torchlight, lifted the bolt cutters to the enclosure door. 'You ken they have polar bears here?'

'For God's sake, you never trust me to do anything.'

The older man waved the bolt cutter. 'Like turning left into that tree, yer mean? If I cut this bloody lock off and a polar bear comes riving out … See these things? The last thing I'll do before the bloody thing eats me will be to cut your balls off with them.'

'I'm telling you it's the right enclosure.'

The lock protested with a metallic squeal as the cutters sliced through. Together the two men heaved the cage into place over the gaping mouth of the enclosure door.

The older man scratched the wool of his chin whiskers and sniffed the air. 'I hope you're right. This cage widnae slow a polar bear down lang.' He pulled his camouflaged jacket closer around his shoulders. 'OK. Give me the bait.'

The young man pulled a packet of meat from inside his jacket and examined it with disgust. 'It stinks.'

The older man took it and tossed it into the cage. 'That's how they like it. Carrion, ken.'

The two men crouched in the darkness behind the Land Rover and waited. Nothing happened for a long time. They grew cold and stiff from waiting and, at length, a thin light began to suffuse the sky. Inside the enclosure the pale light began to pick out bushes and a kind of plank seat suspended by ropes.

The older man rubbed his back and groaned. 'God, I'm stiff.

Nothing's coming.'

The younger man ran his hands over his knees to try and get some warmth. 'Just wait. Patience is what it takes.'

'Something moved!'

'Where?'

The older man stabbed his finger towards the corner of the enclosure. 'By that bush there.'

'I don't see anything.'

'Shh, quiet.'

The younger man sighed. 'There's nothing there. Oh wait ... '

A ghostly shape melted through the darkness and, for a fraction of a second, fierce yellow eyes sparked into life.

The younger man swallowed, fear whirling in his chest. 'You see something?'

A low growl punctuated the darkness.

'Oh Christ.'

WINTER

CHAPTER 1

Winter is a fickle mistress in the mountains of the Scottish Highlands. At times she lures you on with tantalising glimpses of snow-white hills, ranging forever like the gateway to an enchanted frozen world. Then, when the sky fills with grey, saturated clouds and soft rain shrouds the mountains, she becomes mysterious, gentle. She has great beauty, this mistress, and those she touches, those who travel in these high places, are left with an inescapable yearning to return.

Yet for all winter's gifts, for all her power to raise those who love her to unimagined heights, she has a cruel, cold heart, and her anger is something to be feared. On one particularly cold day in February the west coast of Scotland had become a battleground as moisture-laden air racing in across the Atlantic collided with a frigid air mass from the east. The Atlantic gales crashed against the wall of cold, disgorging their moisture in torrents of snow amid howling winds. For two men high in the mountains, a day that had started calm and clear had sprung a storm upon them, and their gentle outing had become a struggle for survival. Winter had them in her grasp and, like the jealous lover she is, was reluctant to let them escape from her embrace.

***

Rory pushed forward into the blizzard. He strained to see the mountain ahead, but the snow blinded him and its jagged fingers pierced his cheeks with an agonising cold. He turned away from the wind and scraped shards of ice from his face. Looking down, he could see nothing but curtains of falling snow that reduced his vision to a monochrome texture and destroyed all sense of depth or perspective. It became impossible to discern the shape of the mountain ahead, let alone find a safe route above the cliffs, but a sixth sense told him that the ridge was narrowing and below him gaped a hungry void. He wondered where Angus had got to.

Ice rattled in Rory's dishevelled ginger beard. He knew now that his handful of years of summer hillwalking and half a dozen weekends in benign weather on the winter hills had been no preparation for what he was facing, in the teeth of a Highland blizzard and with a whole mountain between him and safety. Then he noticed something else: the light was beginning to fade. Soon the short winter day would be over, and he and Angus would be lost in the darkness.

Snowflakes found their way inside his hood and down into his jacket where they melted against his skin, making him shiver as the cold gradually bored its way inside him. As Rory fought to close his jacket, his fingers brushed against the cord around his neck that held his wedding ring, a little band of gold against the warmth of his chest. For a moment he felt Jen standing beside him, her head on his shoulder, just like she always did when the world got too much for him. Then the blizzard roared again and he was back on the mountain – but now he thought he could contain the fear wriggling like an electric worm inside his belly.

Rory turned anxiously and was relieved to see a stocky figure, dressed in an ancient battered cagoule, stumbling towards him through the snow. It was Angus, twenty years or more older and with so much more experience; he would know what to do.

Rory yelled against the wind. 'Angus, which way now?'

Angus, his face wrapped in the hood of his cagoule and his balaclava below that, bent forward to scrape away the accumulated ice. 'Let's awa' back to the bothy.'

Angus spoke calmly, as if suggesting another round of drinks on a Thursday night in the little pub in Inverness where the mountaineering club met. It comforted Rory to have the older man with him. When the mountains confront us with their savagery it is the companionship of others that keeps fear at bay.

Rory saw a reassuring grin flash across Angus's face. 'Ach, it'll be nae bother. Get that gizmo of yours oot. We'll be back in the bothy as soon as you know it.'

Rory visualised a hearth with a roaring fire in the grate and flickering candles on the mantelpiece. He grinned back and pulled the GPS out of his pocket. Then he looked in horror at the screen – it was grey and dead.

'Bloody thing has packed up.'

In that instant the storm seemed to redouble in strength. The world of swirling white grew around Rory until it swallowed him like an ocean with unknown depths.

Angus dropped his rucksack. 'I telt yer, you cannae rely on those bloody things.'

'I put new batteries in it only last week. It should be fine.'

The blank screen stared back at him in mute defiance and he watched Angus pull out his map and try to hang on to it as the blizzard attempted to wrestle it from his hands.

Angus got the map under control and turned to Rory. 'OK, where are we?'

'I was following the arrow,' Rory said, realising that he should have paid more attention to the device while it had been functioning. 'Somewhere here, I think.' He poked at a point on the map with his ice-glazed gloved finger, indicating somewhere high on the ridge.

Angus sniffed and scanned the maelstrom of snow. 'Och well, nae sense in hanging aboot.'

There was a casual tone in his words, as if he were announcing that it was time to head home from the supermarket. But Rory detected a hint of anxiety in the older man's demeanour and realised that Angus was pretending to be calm for his sake.

Rory watched as Angus tried to stare through the snow, looking for some feature that would give some clue to their position. Each time Angus scoured the mountainside for a reference point, the blizzard seemed to delight in increasing its ferocity.

At last he took a bearing with his compass and followed the needle into the whiteness. 'This cannae be far oot.'

Rory knew this could only be a lie, told to bolster his confidence, for there was nothing visible in the heart of the storm to show where they were.

'We've got to find that gap in the cliffs,' Rory yelled.

Angus did not respond, but walked on, guided by the flickering needle. Rory swallowed, gathered his courage and followed.

The slope beneath them grew ever steeper as they began the descent. Rory gripped his ice axe in earnest, driving it hard into the snow with each step, anchoring himself to the mountain.

Sometimes he was completely blinded by the swirling snow, and feared that with the next step there would be nothing but air beneath his boot and he would plunge into the abyss. Although only a few feet away, Angus was sometimes lost for a few seconds in the white-out. Rory forced his trembling legs to move, stamping the spikes of his crampons into the snow and then driving in his ice axe as an anchor before taking the next step. This was the rhythm of his movement, the key to getting down the mountain alive and seeing Jen again. *Don't panic*, he thought, *just keep moving steadily*.

Angus called to him through the blizzard. 'Now just keep your weight over your crampons like I showed you. You'll be fine, yer ken.'

Angus could be patronising, especially when he went into his 'old man of the mountains' mode, dispensing sage advice to anyone who wanted to listen and sometimes to people who didn't. Right now, Rory didn't mind being patronised – in fact he found it rather reassuring. He just hoped the old man could get them off this mountain in one piece.

Suddenly Rory found himself on Angus's heels. The old man had stopped and was peering anxiously below him. 'Rory, I'm nae sure this is the way doon. It's too steep.'

Rory stood shivering. Angus stepped tentatively forward, and in a fraction of a second the ice beneath him sheared away. He fell, and his body spun away towards oblivion.

***

Angus hit the ice-covered slope hard, driving the air from his lungs. His body tumbled with frightening speed as he rolled down into the darkness, bouncing over the short steps of the mountainside. Instinctively he rolled on to his stomach and lifted his heels up to keep his crampons out of the snow, knowing that if they caught in the ice he would begin a deadly cartwheel.

Angus could feel his body moving faster as the slope steepened. *Grab the ice axe*, he thought, and looked up to find the tool already in his hands as if it had leapt there of its own accord.

In the moments that separate life from death, panic kills; only cold reason can push disaster back into its cage. *Not too fast now.* Angus pushed the curved blade into the surface with a gentle but insistent force. Too fast and the axe might grab at the ice and be ripped from his hands as the downward force of his body carried him away into oblivion. The steel bit, and a shower of ice crystals arced into the air. He pushed a little harder and felt the hook of the axe begin to dig in. *Harder now.* Angus drove the pick gradually in to the ice and found himself wrestling in a battle to control it as he fought the pull of gravity.

Perhaps he fell for ten seconds, perhaps thirty; time froze as he hurtled downwards, clinging to the ice axe, his only lifeline. The speed was frightening at first but he felt his body slowing as the ice axe bit into the slope. Then he regained control, felt the axe bite deeper, and his body stopped moving. A wave of relief ran through Angus's body. Only then did he dare to kick the crampon points on the toes of his boots into the ice. He was safe now and lay there for a moment, gasping, letting the air flood back into his lungs and feeling his strength return.

He looked up and not too far away he could see Rory looking down at him, eyes wide with terror.

Oh, for fuck's sake. Angus was angry with himself – he had almost paid the ultimate price for a moment's carelessness. He thumped his axe into the snow above him and with a few grunts and oaths was soon standing beside his companion, puffing from the effort.

Rory was pale and shaking. 'I thought you were gone then.'

Angus grinned back at him. 'Nae worry. Take mare than that to kill me.'

That was a lie, but there are times for telling the truth and sometimes lies are better when the truth cannot be stared in the face.

'I ken where we are now. I mind this place. We're in the top of North-West Gully – I climbed it a few years back. There's nae way doon this way but I can find the descent route from here.'

***

Rory watched as Angus set his compass to a new bearing, and the two men trudged off through the deepening snow in a new direction. The storm attacked again, rocking the climbers in their steps, tugging at them with its teeth in a rage, not wanting to let its quarry escape. The sky bulged, spitting forth great gouts of snow that fought to penetrate hoods and clothing. Rory's world shrank to the next few feet of snow. The battle for survival was the only thing that either of them knew.

Rory plodded after Angus until at last the older man stopped and yelled over the wind.

'This is it. We can get doon here.'

Rory tried to call back but suddenly realised that the effort of kicking his way through the snow had sucked the air from his lungs and left him gasping. 'You sure?'

Angus laughed again but this time it was with relief. 'Aye, I dinnae need one of those bloody gadgets to find my way aboot.'

At that moment the beast in the darkness roared with a rage that would not be denied and hurled both men face down into the snow.

Rory rose, blinded by the flakes, winded from the fall. 'Christ, let's get out of here.'

For half an hour they slogged downwards until, to Rory's relief, the gap in the cliffs they sought opened up before them and they moved more easily on safer ground. As they began to lose height they started to leave the kingdom of the ice monster. Every step took them further away from the wind and soon the furious gusts abated and they had left the storm raging above them. By this point the light had almost gone. Rory was grateful to be following Angus as their head torches painted small, bobbing pools of light before them with only the great empty darkness all around. Rory felt a growing respect for the old man. Despite his fall he had guided them safely down – his knowledge of the mountains had saved them both from disaster.

Experience in the mountains is hard won. You cannot buy it; you will not find it in shiny plastic bags on the shelves of warm, brightly lit outdoor shops. You will find it on a mountainside in the impenetrable darkness, you will find it when you have walked to exhaustion and home is still a long way off. Experience comes from overcoming mistakes – it is dirty, cold and wet, and it comes wrapped in fear.

Rory longed for the warmth and safety of the bothy far below. 'I wonder if Brian's lit the fire.'

'Lit it? He's probably in it.'

There was a calmness in Angus's voice that reassured Rory, and the pair chuckled. Only laughter has the power to push fear back into the dark corner of the mind.

Lower down the mountain they were out of the worst of the weather, and an eerie silence surrounded them. Even their footsteps became silent as boots sank deep into the snow. In this soundless white world, Rory felt as though he were floating down the mountain as everything became soft and insubstantial. Then the snow beneath his feet changed from soft and yielding to firm. He no longer sank into it but found himself walking on the surface. The pair started to move quickly, and though he could not see it Rory sensed the floor of the glen beneath them.

'Hold on.' Angus had stopped and was peering at the surface of the snow. He prodded at the surface, peering at small ripples that had formed, sparkling under the beam of his head torch.

Rory watched the old man, puzzled. 'What is it?'

Angus was silent for a moment. 'Sastrugi,' he said at last. 'Ken ripples are called sastrugi.'

Rory thought this an odd time and place for Angus to suddenly start giving a lecture on the nature of snow. 'So?' he grunted, patience wearing thin.

Angus wiped the melting snow from his face with the back of his sleeve. 'It means this snow was deposited by the wind, yer ken.'

'Oh right.'

Rory took a step down the slope but Angus held up his hand to stop him.

'It's what they call "windslab".' Angus turned, his beard

jewelled with melting snow. 'It's avalanche-prone, risky stuff on this slope. And it's been falling all day.'

Rory's mind filled with images of great blocks of snow tumbling down the mountain, carrying the hillside with them. His mind went back to a lecture on avalanches he'd been to at Glenmore Lodge, the mountain training centre in the Cairngorms. He remembered an earnest young man with 'No Fear' on his T-shirt talking about layers of snow, and bonding, and something called 'graupel'. But Rory had dozed at the time and dreamed of being beside a bothy fire far away from the lecture hall. Now he wished he'd paid a bit more attention.

'We'll have to go around it, then.'

***

Angus thought of the long miles of walking it would take to avoid the slope, of the hours plodding through soft snow, and he remembered something too – he remembered that he was fifty-five not twenty-five. Thirty years ago he would have grunted and set off back over the hill, but not now. He knew he had to take the chance and go down. Round was not an option. Angus hated to admit that age had taken its toll and he was not as fit as he once was. Old mountaineers can fool themselves that they are still young, sometimes they even fool other people, but they don't fool the mountains. The hills are patient. They wait for old men and women to make mistakes and remind them of how many years have passed and what dues are owed. Angus felt the ache in his legs, felt how heavy they were.

'I'm too tired,' Angus admitted. 'Yer go around it. I'll have to chance it.'

Rory didn't budge. Perhaps he was unwilling to leave his companion. 'You're supposed to swim if you get caught in an avalanche, aren't you?'

'Go round, Rory, there's nae point in you taking the chance.' Angus was pleading with him now, not wanting to take the young man into danger, but he could see that Rory's mind was already made up.

Rory shrugged, shifted his rucksack higher on to his shoulders, and set off down the slope. 'Come on, you old fart. If we don't get there soon that bastard Brian will have burned all the coal.'

He would never admit it but part of Angus was pleased that Rory had not left him. He did not want to be alone in that place, but he was also touched that the young man would so easily risk his life to remain at his side. Mountain law is simple. You never leave a companion, even if you have to die together.

Now the snow began falling heavily once again. The sky and the mountainside lost their separation; there was no horizon, no sky, no earth, just a huge, all-consuming whiteness. They walked in silence, tense and afraid, listening for the slightest sound that the snow was about to slide. Angus's back ached from the hours of walking, and his brain wanted to shut down, to sleep, to be somewhere far away from this frigid nightmare. They were moving quickly now. The snow found a home for each pace they took; there was no need to worry about finding a secure place to step, no risk of turning an ankle. Sometimes it was hard to believe they were moving at all. They walked but nothing changed, no boulders passed, and all perspective vanished into the limitless white.

Angus stopped and rummaged in a pocket. He produced a bar of dark chocolate, broke off a piece and offered some to Rory, who took it gratefully. 'Careful now, it's bad for yer, all that sugar.'

'Fuck it, I'll be lucky to get off this mountain alive.'

After twenty minutes of stepping into nothing a huge dark shape heaved into view, like a ship sailing through a sea of mist. It was a rock formation Angus knew well.

'Thank God, we're almost doon. Let's get on to that rock ridge. I dinnae like this snow.'

Lured by the security of solid ground, they hurried towards the rocks. Angus hauled himself up on to the rock and turned to Rory, who came towards him like a ghostly snowman. Then there was a crack, a sound that split the air, and Rory was gone.

***

In an instant Rory's world became a great churning mass of cold oblivion. It spun and rolled until down and up no longer had meaning. His torch was ripped from his head; he saw it bounce away until it vanished and the dark swallowed it. He clawed to escape the jaws of crushing snow around him, flailing with all his might to stay on top of the surface, kicking in desperation, all the time expecting the darkness beneath to suck him down and suffocate him.

The rock, the rock. Make the rock. The words ran through his mind like a mantra. He lunged, and his hand caught something solid, something not sliding down the mountain. *The rock! It's the rock!* He pulled and kicked while the mountain shook and rumbled like a growling bear.

Suddenly he was out, gulping lungfuls of air in the darkness. Feeling solid ground beneath his feet for the first time in hours, there he lay, too exhausted to move, grateful for the jagged boulders that stabbed into his back. Far below the sound of the avalanche retreated and the beast was gone. A light bobbed down from above through the blackness and soon Angus stood over him, dazzling him with the torch.

'Christ, are yer all right?'

Rory eased himself to his feet. His legs shook but as far as he could tell he just had a few bruises.

'Aye, I think so. I don't want to play this game any more. Let's go home.'

Angus strode on but the avalanche had knocked all of Rory's energy out of him. His body ached where the snow had almost crushed him or twisted his limbs at odd angles. He had felt no pain in the moments after he had crawled from the snow, but now every movement brought fresh agony and his body screamed for sleep.

By the time they reached the floor of the glen, the snow was turning to slush and wet sleet spattered their jackets. They walked in silence, too tired to talk, and only the need to reach the shelter of the bothy propelled them forward. They scoured the blackness for signs of a light.

'Four hundred and twenty-five, four hundred and twenty-six.' Angus counted off the steps, the only way to measure distance in the featureless night.

Rory stopped, the pain in his back increasing and his bruises from the fall aching more and more. 'We've passed it. We must have.'

'Nae, not yet, we've nae come far enough.'

'We'll never find it in this bloody snow.' Despair overwhelmed Rory and he tore off his rucksack and sat down on a boulder. 'Angus, I'll have to rest for a moment.'

For twenty minutes Angus left Rory seated on the rock and scoured the area for the bothy. After what seemed like forever, Rory watched Angus's torch beam come dancing through the blackness towards him. The older man did not need to speak. The look on his face told Rory the little shelter had eluded him.

'Come on, Rory, it cannae be far now. Just another twenty minutes. We'll find it.'

Rory staggered to his feet, wincing from the pain; even standing felt like a struggle. The rain had soaked them both through. A night out in this weather could prove fatal. They needed the bothy more than ever.

Rory lurched after Angus, legs crumbling with every step. He no longer scanned the blackness for any sign of light that might give away the position of the bothy – he simply followed the old man and trusted his ability to guide them home. But the path was buried under melting snow and half an hour later Rory's legs gave way, pitching him forward into the slush. The avalanche, the cold and the darkness had taken their toll.

Moments later he felt Angus lifting him.

'I've got to rest,' Rory murmured.

Then the hallucinations began. He saw a light coming towards them out of the dark. It drew closer and closer, then an alien-like figure emerged from the glow and began to slowly beckon.

'Oh, there you are,' came a voice from out of the night. Beneath a fading head torch beam an emaciated figure dangled.

'Brian!' Angus called to the figure, his voice betraying amazement and disbelief. 'What are yer doing oot here?'

'Looking for you,' Brian said casually, as if he'd been waiting to meet them in a pub or on a park bench.

His features were gaunt and drawn. He looked as if he hadn't had a decent meal in years. Around his bony shoulders hung an ancient orange cagoule; it was saturated and held together by odd patches of material. Many years ago it might have been able to withstand a Highland storm but now it was worn so thin even the briefest shower would surely penetrate it. Brian had obviously been out for hours and looked as worn out as his gear. His grey hair hung limp, dripping water across a face that had been beaten into creased leather by years out in mountain storms. Another man might have been pleased to see his companions, might have extended his hand in greeting – smiled perhaps, or even thrown his arms about them – but Brian was not given to outward displays of emotion. On bothy nights, when the others grew boisterous after surfeits of whisky and beer, and songs were sung and tales told, Brian would be sitting in a corner, barely noticeable, like a scrawny spectre at the feast.

'How far is the bothy?' Rory demanded, too exhausted for pleasantries.

Brian turned and pointed. 'There it is.'

Rory followed Brian's torch beam. At first he saw nothing but as he stared into the blackness a ghostly shape gradually formed amongst the falling snow. Rory could see how they had struggled to find it. The bothy was a simple construction, made from the rough stone of the glen itself, so covered by lichen and moss that it looked to have grown from the ground rather than having been built by human hands. From only a few footsteps away the bothy appeared to be a boulder or a stand of trees.

Rory bent his head in the low doorway and stepped inside. He stood for a second, dripping water on the floor as he took in the scene, thinking how wonderful it was to be out of the storm at last. It was as if he had stepped back to a time before electricity, before running water and before we complicated life with the trappings of luxury and convenience. In the semi-dark he made out a few embers of glowing coal in the dying fire. On the mantelpiece above the rough stone hearth, candles flickered from long-dry wine bottles. The room was lined with wood and boasted a few meagre bits of old furniture: a bench, a rickety table and a couple of chairs. In one corner two muffled bodies lay in sleeping bags. They stirred for a few moments on hearing the three men enter the bothy only to return, in seconds, to their slumber. The air was thick with the smells of drying socks, sweaty bodies, woodsmoke and recently eaten meals. If he had walked in off the street the cacophony of odours would have repelled Rory, but now, after hours out in the hostile blizzard, he thought he had walked into paradise.

Brian got to work as Rory and Angus climbed out of their dripping hill clothes. Brian still used much of the equipment he'd used for the last forty years. He had never made the transition from paraffin to gas. Now he produced his ancient brass Primus stove and went through a well-rehearsed ritual, lighting a small quantity of methylated spirits in the well beneath the burner to bring the stove to its operating temperature. He pumped just the right amount of pressure into the paraffin tank and waited for the spirit's blue flame to heat the burner until it would light. The instant the stove caught, it filled the bothy with the smell of burning paraffin and a delicious purring sound: the sound of warmth and food and comfort. Rory watched Brian work with a silent, precise efficiency. Once he had the kettle on he set out two plastic mugs, one for Angus and one for Rory – nothing for himself until the others were fed. Then Brian produced some coal he had saved from the fire for their return. Wordlessly he took their dripping outdoor clothes and hung them over a makeshift pully as the coal fire spluttered into life. He filled the mountaineers with tea and soup until they were both warm and the feeling had returned to their toes. Only then did he eat a little himself.

It was well past midnight when Angus and Rory sat beside each other on the crude wooden bench before the dwindling fire. Angus puffed on his pipe as he watched the flames and Rory cradled his beer can. In the corner, Brian snored loudly, but for the pair it would take time for the adrenaline to leave their bloodstreams.

Rory sniffed. 'He never says much, Brian, does he?'

Angus inhaled the aroma of his favourite Irish Cask and blew a smoke ring out towards the chimney. 'Suppose not. Taciturn is the word, I believe.'

Rory winced as he felt his ribs. 'You mean miserable.'

Angus gave Rory a reproachful glance through the pipe smoke. 'He searched for us for three hours ken when we didnae come back. Yer've got to give him that.'

Rory glanced again at the sleeping figure and then looked down at the dwindling dregs of his beer. 'You're right. You know what they say, "If you can't say something nice, say nothing at all".'

Angus thought for a moment. 'Is that Descartes?'

'No, Thumper in the film *Bambi*.'

'Oh aye, the rabbit.'

'That's him.' He coughed and clutched at his ribs again. 'So sad, that film. When Bambi's mother gets shot by the hunters.'

Angus smiled and nodded, recalling a distant memory. 'I mind when I went to see that film with my mother. Must have been six. There was nae a dry eye in the cinema. We went to the pictures every Saturday back then.'

Rory took a sip of his beer. 'I can't remember the last time me and Jen went to the cinema. We just get a DVD out. Romcoms she likes.'

'Changed days.' Angus sighed. 'When I was a loon there were over thirty cinemas in Aberdeen. The Gaumont, the Astoria. All plush carpet. Gone now.'

'That's what turned me vegan, *Bambi*. I still can't see why anyone would kill anything for pleasure. Let alone a deer.'

'Didn't yer no say there were too many deer?' Angus tapped his pipe out on the hearth as the fire began to die.

'I'm not against controlling them. I just can't see why anyone would take pleasure in killing anything.'

Angus began to unroll his sleeping bag and rolled his fleece into a pillow. 'Where would the Highland economy be without deerstalking? Would put a lot of folk oot of a job.'

Rory opened his mouth to argue but only a yawn escaped his lips. It was late and sleep was more important than discussion. He stood, wearily, and found another bruise on his thigh in the process.

Angus blew out the candles and they both climbed into their sleeping bags. The room fell into darkness, lit only by the odd flicker from the dying fire.

As they settled down to sleep on the wooden floor of the bothy, Rory spoke. 'That was close today.'

Angus grunted and wriggled deeper into his sleeping bag. 'Aye, bloody close. Yer dinnae see snow like that these days. Winters aren't like they used to be.'

Rory yawned, his body craving sleep. 'Global warming. We're killing this planet. Suffocating it bit by bit.'

'Aye, I used to love the snow. The cold and the freshness of it. Jesus, I'm tired.'

Rory laughed. 'I don't know why we do this.'

Angus replied in the mumble of one whose brain is already wandering in the land of dreams. 'It's either this or watch *Strictly Bloody Fools Dancing*. Either way you wind up dead.'

Rory closed his eyes and tried to sleep, yet despite the exhaustion his brain was still working to understand the events of the day. Images of the storm played on the cinema screen of his eyelids. Horror filled him as he watched the ice give way beneath Angus and saw him fall towards the yawning cliffs. Worst of all, his mind took him into the jaws of the avalanche. He felt the darkness, the crushing weight and, as sleep drew closer, he reached out for the rocks that had hauled him from death.

In his mind Rory was back on the mountain tumbling over and over as the avalanche tried to swallow him. The scene played on repeat in front of his closed eyelids until, at last, his brain came to terms with the horror and let him drift off to sleep.

Beyond the bothy door, while the men dreamed inside, a stag ambled past in the moonlight. The snow had ceased falling and the glen was still and empty as the sky cleared and the frost descended. The stag froze for a second, catching the scent of humans and fire as he passed the bothy window. He did not know that next October he would linger too long on a hillock and a bullet would tear through his heart and leave him bleeding in the heather.

Far above the bothy, the storm's rage spent, the mountain-tops were silent once more. Here and there a lazy wind tossed handfuls of spindrift into the air where the moonlight caught them for the briefest of moments. Only a few faint footprints in the snow remained of the climbers' struggle. All other signs had faded like the memory of a bad dream, and soon they too would sink into the night.

CHAPTER 2

The old ghillie was quietly daydreaming in the passenger seat when the Land Rover screeched to a halt, sending his shoulder into the seatbelt with a bruising jolt so sudden that his deerstalker hat flew off. The driver, a large young man in a tweed jacket, mumbled something then picked up his binoculars and peered out of the car.

'Och, Hamish, what are youse doing?' the passenger said, retrieving his deerstalker from the footwell and cramming it on top of his shock of grey hair.

Hamish turned, his face round and flushed, and yelled, 'Walkers, Donald! I said walkers!'

Donald was tired of being shouted at by the young keeper. 'There's nae need to shoot. I can hear ye perfectly well.'

Hamish rolled his eyes, grunted, and thrust the binoculars at Donald. The old keeper climbed stiffly out of the Land Rover and scanned the hillside. The vehicle had come to a halt outside the main entrance to Castle Purdey. Behind him loomed a great dark Gothic edifice, with turrets at its corners and castellated walls where fearsome gargoyles stared out with blind stone eyes. In front of him a vast lawn stretched away, circled by the gravel drive where the Land Rover stood, its engine throbbing. In the middle of the lawn there was a fountain, a naked cherub boy at its centre endlessly urinating into the pool below. The sound of the tinkling water created an instant echoing response from Donald's bladder and he cursed the infirmities of age. Across the parkland, beyond where heavy beef cattle grazed, was a tall dark wood of old oak trees; beyond that a path led out on to the open hillside towards the mountains with a fringe of snow far above. He searched with the glasses across the pasture up through the wood to the hillside above but saw no walkers.

'Where are they, Hamish?'

Hamish sighed, and that was another thing the young man did that irritated Donald: he sighed a lot.

'There, can you not see them?' Hamish pointed in frustration. 'Near the wood.'

Donald saw two figures with large packs heading slowly up towards the trees. They were some distance away so he could make out few details. A tall, thin man was in the lead, followed a few yards behind by a shorter, stocky figure who moved a little more slowly.

'Och aye, I have them now.'

The sky was dull grey and a pair of ravens cawed their way into the air as the walkers entered the wood.

Hamish grunted and lowered the binoculars. 'Friday night. They're heading up to spend the weekend in the bothy. All the way up yonder just to spend a few nights in a bloody ruin of a place. Hillwalkers! Townies playing at being country folk. Thinking that they can tell us what to do on the land.'

A black Daimler shot through the gates and slewed to a halt beside the Land Rover. The door flew open and a tweed-clad figure bundled himself out on to the driveway. Lord Purdey stood there for a moment, puffing in indignation. He was stocky and bald with the type of moustache normally only worn by wartime fighter pilots in B-movies. He thrust his fists into his sides and stood for a moment watching the walkers make steady progress up the track.

'Again! What is it with these people?' His accent was crisp, expensive.

Hamish shook his head sadly. 'Friday night. Always bad for them is Fridays.'

The bald man thrust his head through the open window of the Land Rover and pressed his face so close to Hamish that their noses almost touched.

'They won't *find* anything, will they?' This he whispered, although only the keepers were close enough to hear him. If he had wished, he could have used a megaphone and still maintained the same degree of confidentiality.

Donald shook his head. 'Nae sir, nae sir.' There was a pause, during which time it was just possible to hear the cogs in Donald's head turning.

It was Hamish who said what the older ghillie was thinking. 'As long as they don't wander up on to the Black Moor, that is.'

Donald remained silent but nodded sagely.

Lord Purdey glared, his moustache bristling. 'Now look, we can't afford any ... ' He searched for the right word. '*Embarrassments*, not with that RSPB arse sticking his nose in.'

Donald nodded and Hamish did the same.

Purdey fell silent and paced the driveway deep in thought. At last he turned and spoke pensively. 'Donald, you remember, in my father's day, we could have seen those map-and-compass chaps off. But now ... ' He sighed.

'Different days, sir. Right tae roam legislation.'

Purdey nodded slowly. 'Right to wander about and make a bloody nuisance of yourself. Makes you wonder whose land this is.'

'Och aye, that it does, sir,' Donald said, binoculars focused on the two figures. 'Suppose we could tell them some pish, like we're shooting up there.'

Purdey sniffed, thrust his hands into the pockets of his yellow corduroy trousers, and began pacing again. 'Shooting, yes, good idea. Tell them that.'

Hamish climbed out of the Land Rover, squeezing his bulk through the door. 'I wish we could shoot hillwalkers. If this was America we could. That Donald Trump, that's who we need.'

'Och now, I don't know about that. If this wis America they might fire back anaw,' Donald added with a laugh.

Purdey smiled thinly and chuckled with them. 'I'm not sure about Trump. He'd turn Scotland into a golf course and put a wall across the border.'

'Would that be to keep the English out or the Scots in, do you think?' Donald said.

Purdey's smile vanished. 'Head up there and tell those bloody map-and-compass Johnnies to clear off.'

***

Rory could feel the sweat condensing against his back as he climbed the steep track that wound up through the woods and out on to the open hillside. Out of the shelter of the trees the wind picked up and he was glad to feel the cold February air against his cheek as he waited for Angus to catch him. Once he had stepped out of the trees into the wind, with the open moorland before him and the snow-capped hills against the horizon, he knew he was in the mountains again. The wind was alive with the scent of heather and moor and the taint of distant snow, and it embraced him like a lost friend. The only sounds were the babble of a burn close by and the sighing of the trees as restless wind pushed its way through. The factory that enslaved him five days a week was far away; now he didn't have to take orders or follow ceaseless rules. Now the only things that mattered were the earth beneath his boots and the endless open sky. Now he could breathe. Now he was free.

The light was fading and the outlines of the hills softening against the bullet-grey sky when Angus arrived, his beard glistening with condensation and his face dripping with sweat.

'Jesus, that hill does nae get any easier, does it?'

Rory was used to waiting for Angus. He'd waited for him on mountain passes and summits across the Highlands, but he couldn't resist the opportunity to tease him.

'It's that pipe you smoke. No wonder you can't get up hills.'

Angus grimaced in mock pain. 'You're starting tae sound like Laura. Can't a man have one small pleasure in life? Albert Einstein himself said that pipe smoking contributes tae a calm and objective judgement in all human affairs.'

Rory shook his head in silent disbelief, popping a handful of raisins into his mouth. He offered some to Angus, who declined them.

'I suppose it's all relative.'

Angus stood for a moment scratching at his grey beard, a confused expression on his face. 'Was that an attempt at humour?'

Rory chewed slowly on the raisins, trying to decide if he could rescue the joke, but eventually he decided it was beyond help. 'Probably not.'

By now Rory had cooled down and was beginning to feel the windchill. He zipped up his jacket. Rory didn't set much store in material things. Expensive cars or flashy watches left him cold. What he really valued was a day in the hills with clear views or a night in a remote bothy beside a warm fire. He had only discovered hillwalking and climbing a handful of years ago – a novice compared to Angus – but he was already learning the value of good equipment. Out on the hills an expensive watch wouldn't save your life, but a decent jacket just might. The trouble was, his wages at the factory didn't stretch far enough.

Rory was shouldering his pack when he glanced down the hill and spotted a Land Rover making its way towards them. He was already angry before the vehicle got to them. The enemy was coming – the foot soldiers of the landowners, who tried to keep these estates as private empires where they could do what they wanted, destroying the wildlife and rendering the land barren. They met such men (and they were always men) over and over again in their wanderings. Angus always respected their authority but Rory had no time for them.

The Land Rover came to a halt. A big ginger ghillie, in the blood-sports uniform of tweed, leaned through the window.

'Where are you off to, boys?'

'What's it to you?' Rory snapped. 'We've every right to be here.'

The young keeper glowered at Rory and stepped slowly from the vehicle until he stood towering over him. 'You keep off the

Black Moor, we're shooting there.'

The two faced each other like dogs reluctant to give ground.

Angus put his hand on Rory's shoulder and spoke quietly to the keeper. 'We're awa' tae Glen bothy for a couple of nights.'

Rory ignored his friend and was in no mood to back down. He couldn't see why they should be challenged when all they were doing was walking where they had a right to go.

'We'll go where we like.'

An older ghillie called from his seat in the Land Rover to the increasingly furious estate worker, who had drawn himself up to his full height. 'Hamish, hawd on, come away in now.' Then he spoke to Angus gently, holding his palms out and offering a rueful smile as if keen to defuse the situation. 'Nae problem if youse heading for the Glen bothy. If you could jist mind tae keep off the Black Moor. It's for safety.'

'We're quite happy tae go a different way if it makes things easier,' Angus said. 'As long as we get tae the bothy.'

'Just don't go on the Black Moor,' Hamish insisted, his face still red.

Rory felt his hackles rise again. 'Why, what's on the Black Moor?'

'I told you, we're shooting there. Right, Donald?'

The older keeper began to nod but Rory butted in. 'What, tonight? It's almost dark.'

'*Vermin control*,' Hamish said through clenched teeth.

Donald called to the young man, insistent this time. 'Hamish, awa' back into the car now. We don't want tae start a rammy.'

Angus made another attempt to broker peace. 'We'll jist head tae the bothy now. We're nae after going on to the moor.'

Hamish took one last hard look at the two hillwalkers and turned back towards the vehicle.

Rory, as angry and red in the face as Hamish, called after him. 'And what do you call vermin? Hen harriers, red kites, golden eagles.'

'Youse'll find nothing fly up here now,' Donald replied, tone still quiet and reasonable. 'Those birds is protected, legal like.'

'I wonder what we'd find if we took a walk on the Black Moor,' Rory said.

Hamish took a step back towards them.

Donald's patience finally snapped. 'Hamish! Get in the feckin' car.'

The ginger gamekeeper snarled and shrugged and climbed back into the four-wheel drive. There was a crashing of gears and the Land Rover careered off across the hillside, leaving the walkers standing in a cloud of exhaust fumes.

'Purdey's men,' Rory said in disgust as he watched them drive away across the moor.

Angus patted Rory on the shoulder, as if to calm him. 'Aye well, we all have to make a living. Live and let live, yer ken, that's what I say.'

Rory set off up the track, calling back as he walked: 'That's just the problem. They live to kill, that's the name of their game.'

***

They walked on in silence for a while, each lost in his own thoughts. The path skirted the Black Moor and then turned off Purdey's estate and on to the Muir estate. A fence clearly marked the boundary although it would have been easy to distinguish them without it. The Purdey side of the fence was open, dead and empty moorland, while the Muir side was bursting with small trees and returning life as his estate encouraged regeneration.

Rory was holding the gate open for Angus when his eyes caught a white flicker in the heather 200 yards away. At first he thought he was mistaken but then a white wing rose from amongst the brown. A gull, he thought – but no, not here in the mountains.

Rory fumbled for his binoculars. 'Angus! Did you see that?'

Angus followed Rory's gaze and the bird rose a little.

Rory caught a glimpse of black on the tips of the white wings and a surge of excitement ran through him. 'Hen harrier, get down, quick!'

They both ducked into the bushes and Rory fixed his binoculars on the bird. As if in answer it soared into the sky and Rory watched in awe as the sleek-bodied creature leapt upwards, claiming the air as its own. Now it moved too quickly for him to follow with the glasses and he dropped them just as the harrier turned on to its back and plummeted downwards so fast it seemed it must dash itself to death on the hard earth. When it seemed certain to crash the white wings twisted and once again the creature soared up into the sky.

Rory turned to Angus in delight. 'It's a male hen harrier. I've seen them before, but never like this.'

Angus watched the aerobatic display. Rory knew that the older man was a hillwalker to his core, with only a passing interest in the natural world, yet fascination was written on his face. 'What's it doing?'

'There must be a female somewhere. He's showing her how

good he is.' Rory's voice bubbled with excitement as he watched the bird tumble through the air over and over again.

'I've nivver seen the like. Look at the way it moves in the air.'

As they watched the harrier's dance they were spellbound in wonder and it was as if time had frozen around them.

To see a wild creature is to gaze at the heart of nature. Our land has been flayed, stripped bare until only the bleached white bones remain. The great forests of impenetrable secrets are no more. Our world has been levelled and buried beneath concrete and steel. Even what seems wild is an illusion – our hills are bare, wet deserts kept barren by the gnawing teeth of deer and sheep, relics of the Victorian era. We have no wilderness. We have buried the soul of the land beneath roads, factories and houses. What few wild creatures are left cling, shipwrecked, to the edges of our world.

Through all this destruction something has remained. For those who take the time to look, or are lucky enough to wander the empty places, there are glimpses of wildness to be seen. It is in the tail flick of an otter in the pool of evening, it is in the silent sweep of a barn owl as it heads off to hunt in the failing light, it is in the bright eyes of a pine marten fleeing into the forest, and it is in the dance of the harrier. We come from the land and the memory of it remains in all of us.

They watched for ten minutes until at last the female took to the air. She was larger than the male, her plumage a dull brown with ring markings on her tail. They were together in the air for a few moments and then she turned and the pair soared away over the near horizon.

Angus watched the birds until they vanished from sight,

and the two men sat in silence for a moment. Rory packed away his binoculars and was pleased to see the impression the birds' remarkable flight had made on Angus.

'Never in my life seen that,' Rory said.

Angus nodded. 'Nae, me neither. I didnae ken such things existed.'

Rory smiled at the old man and laughed. 'We'll make a twitcher out of you yet.'

Angus pushed his way through the gate and headed off towards the wood. Rory stood for a few moments, searching the skies for another glimpse of the bird, and his mind filled with images of the hen harriers.

'That was something. That was really something,' he murmured to himself. 'That was the sky dance.'

***

As Rory stepped through the low door and into the bothy he felt as if he had slipped on an old, well-worn overcoat. The familiar smell of woodsmoke greeted him like an old friend and he thought how good it was to be at ease with the world.

Rory lit the candles and dumped a small but heavy carrier bag of coal beside the hearth. 'Shall I do the fire?'

Angus nodded, picked up an empty plastic water container and stepped outside, heading out into the dark glen for the burn close by. There was little need for conversation as the pair settled into a familiar routine: one would light the fire and the other fetch water.

Rory cleaned the old ash out of the hearth. Plumes of dust rose in the candlelight as he shovelled the cold embers into a battered steel bucket. He layered some coal on the cracked grate, then added a few foil-wrapped firelighters, followed by some sticks of kindling he'd brought with him. Finally he topped off the arrangement with a layer of small pieces of coal. Rory always looked forward to lighting the bothy fire. It was the moment that transformed the cold, dark, abandoned cottage into a home. Over the past few years he'd learned a lot about bothy fires. He'd found out how unpredictable and temperamental they were. Sometimes they would roar into life; at other times they would barely smoulder and would sit dark and miserable for hours, like a sulking child. On his early bothy visits there had been times when the fire had refused to light and he had spent long, cold nights staring at a black hearth.

It was Angus who had shown him how to make a fire, taught him when to blow on it and when to leave it be to build heat. Rory had learned the dark art of bothy fire-making well, and now thought that he had surpassed his master – but, of course, Angus would never admit that.

The door burst open and Angus returned from the burn, breathing hard from the effort of carrying the sloshing water container. As Rory applied his match to the firelighter Angus came over to inspect his work.

'Aye, nae bad. Perhaps another dod of coal.' Angus could never resist making some minor suggestion.

A couple of years ago, when they began walking together, Rory would have been irritated by the old man's presumption that his seniority gave him the right to pontificate about all aspects of hill life, but now their friendship had grown he was simply amused.

'It'll be fine, just you wait and see.'

The fire had begun to flicker into life by the time Angus came to the table and set down a steaming plate of instant mashed potato and stewed steak. 'Ah, I'm ready for this.'

Rory looked over disapprovingly. 'Charred corpse, that's what that is.' Rory was eating a bowl of chickpea curry.

Angus paused between spoonfuls of gravy. 'I believe it's Aberdeen Angus.'

'No animals were harmed in the making of this curry,' Rory said.

'Chickpea curry. Hope yer'll be sleeping outside. I don't fancy getting blown oot of the bothy.'

Rory finished his meal and leaned back, content. 'She makes a good curry, does my Jen. A little natural gas does you no harm.'

Angus shook his head sadly. 'All that prime Aberdeen Angus oot there and yer eating peas. Tae think I'd be sharing a bothy with a vegan.'

'Couscous too. I'm well looked after.'

Angus snorted. 'Couscous? What's couscous?' Without waiting for a reply he rose just as the kettle came to the boil. 'Tea?'

Rory held up his mug. 'Aye, thanks.'

Rory settled himself on the bench beside Angus and they sipped their tea and watched as the bright flames began to dance in the hearth. 'There's the bothy telly on.'

'Aye, it's only got one channel but it's a braw show.'

The fire had worked its magic. The cold, rough dwelling they had entered was transformed by the firelight into a sanctuary not only from the cold wind prowling outside but also from the world itself. To spend a night beside a bothy fire is to journey into dreams.

Rory blew the steam from his mug and took a swig of the brew. The tea was strong and hot and he relished the taste even though it seemed to him that powdered milk never managed to pull off its impersonation of the real thing.

'I love a real fire. Me and Jen have just got central heating. When we get our place on the island, I'm getting a wood burner.'

They sat in silence for a while, Angus priming his pipe with his favourite Irish Cask tobacco and Rory swirling the tea around his mug.

Angus lit his pipe and turned to Rory as if something troubled him. 'See what yer told me aboot those ghillies killing birds of prey, eagles and the like – is that true?'

Rory nodded slowly, and a sadness crossed his face. 'Yeah, of course it is. Didn't you know that?'

'Nae. But it's illegal. Those are protected birds, like the ghillie said.'

'Of course it's illegal, but who the hell knows what goes on up in these hills? There's no one here most of the time.'

'Cannae see why they'd want tae do it. What's in it for them?'

Rory was surprised by Angus's ignorance. 'Open your eyes, man. Where we met the keepers, that's grouse moor. You saw the butts, didn't you?'

Angus blew a smoke ring, deep in thought. 'Those wee shelters they hide behind to shoot birds? Aye, there's hundreds of them.'

'You know what they do?'

'Aye, aye. Beaters drive the birds towards them and blokes with guns blast them oot of the air. I widnae fancy it myself but each to their own. If they want to shoot grouse I dinnae see the problem.'

Rory rubbed his temples and felt a wave of despair overwhelm him. 'For Christ's sake, Angus. It's not the shooting of grouse that's the problem. It's the way a grouse moor has to be managed. They eliminate predators so that there is an artificially high number of grouse on the moors, so there's plenty to shoot. The problem is that the place then becomes a magnet for raptors. So there's no way you can manage a grouse moor without killing birds of prey.'

'Like that hen harrier we saw?'

'It's them it hits hardest. Hen harriers range far and wide, so even if they don't live on a grouse moor, they get drawn to them. There's only 400 breeding pairs in Scotland when there should be 5,000. It's worse in England – only a handful.'

The wooden door of the bothy rattled and opened slowly. Both men turned to see who was about to enter. A thin figure walked through.

'Oh, it's yourself, Brian,' Angus said.

Brian stood before them, wearing tattered woollen breeches and an old pullover that was little more than a series of holes held together with darning twine.

'Yes,' came the faint reply. 'It's me.'

Brian was now in his early sixties and had spent most of his life working as an accounts clerk for the NHS. In truth it was clear to everyone that he had missed his vocation. If ever a person had a gift for a role, it was Brian, who should have been an undertaker, for an atmosphere of deep, impenetrable sadness surrounded him. Anyone seeing him following a hearse, or perhaps observing his skeletal frame staring vacantly into a grave, would know that whoever the dear departed was, their demise was greatly regretted by all concerned.

Silence deepened in the room while Rory struggled to find something to say, Angus having already given up any attempt at conversation with Brian. Eventually he resorted to that most British of conversational devices, the weather.

'Rain kept off, then.'

Brian nodded and began to unpack his rucksack. His supplies for bothy trips were, as ever, meagre. He slowly unpacked a cheese sandwich and began to chew on it.

Rory attempted conversation once more. 'Got a few hills planned this weekend, Brian?'

There was a prolonged pause while Brian swallowed a lump of cheddar and best white. 'Maybe.'

A grey mist of gloom seeped out of Brian and slowly filled the bothy. Soon the only sound was a steady drip of rainwater coming through the roof. Before Brian arrived, the drip had seemed inconsequential and Rory and Angus had simply placed an old, rusting pan beneath it and got on with their lives. Now, however, the globules of water that fell from the ceiling took on monumental significance. Three pairs of eyes were transfixed on another silver drop of moisture slowly forming above them. The atmosphere grew tense as the drop grew in size and seemed for a few seconds to defy gravity before at last plunging into the saucepan below with a resounding splosh. They watched it fall and then looked up to the spot on the ceiling as the drip formed once more.

Brian sniffed, his eyes bright with excitement. 'Fifteen

seconds that one took to fall! Two seconds less than the last one.'

Some things are inevitable. Rory did not yet know it, but it was simply a matter of time before, confined for hours in that tiny, remote shelter while Brian timed the water drops, he began to contemplate murder.

**

A handful of miles away, in the great crenellated edifice that was Castle Purdey, Tabatha Purdey strode along the wood-panelled corridor of the mansion she shared with her husband, Charles. She pulled her shawl tighter around her shoulders, feeling the chill as she approached the door of the great hall itself. The 'big house', as it was known locally, boasted three large entertaining rooms, including one that was known as the ballroom; and it had eleven bedrooms, not counting the servants' quarters. It also had, for no reason that was apparent to Tabatha, two medieval-styled turrets complete with gargoyles who stared out across the Purdey estate like ugly, startled five-year-olds. What the house did not boast was central heating that could cope with the Highland weather. Somewhere down in the cellar, a great beast of a boiler did its best to consume the entire oil output of the North Sea every month while only raising the radiators to slightly above freezing. The house was full of draughts, doors that didn't fit and windows that rattled.

For the last twenty years, Tabatha – now in her fifties but still as strong as a rugby fullback – had been trying to persuade Charles, almost twelve years her senior, to move to London where it might be possible to spend an entire evening without experiencing frostbite or hypothermia. Charles, however, had an unerring loyalty to this ancient pile where seven generations of Purdeys had shivered through 300 years of Highland winters. She paused outside the dining room door and allowed the four little dachshunds which dogged her every step to catch up. Through the thick oak door she could hear her husband's voice, doubtless recounting the shooting of some unfortunate beast.

Tabatha ran her eyes across the rows of stags' heads forlornly staring out from the walls where they hung in mute accusation of their killers. Almost as numerous as these poor animals were the portraits of generations of Purdeys, whose expressions seemed just as forlorn as the dead deer. They all had her husband's solid round face and stocky frame. There were Purdeys on horses, Purdeys in armour, Purdeys smoking and casually leaning against enormous cars. For some reason she couldn't quite put her finger on, not one of these dead Purdeys looked as though, all things considered, they'd rather be somewhere else.

She looked down at one of the dachshunds sniffing loudly at the dining room door. 'Queen Victoria has a lot to answer for, Toby.' The dog looked up at her and wagged its tail with that sort of uncomprehending cheerfulness that only small dogs seem to possess. 'I'd like to give that old bird a piece of my mind.'

Tabatha reflected that it was Victoria's passion for a mythical kilt-wearing, bagpipe-playing, hunting-and-shooting Scottish Highlands that had lumbered Tabatha (and a great deal more folk who should have known better) with all these stags' heads, tartan carpets and kilted halfwit husbands who would rather die of cold than abandon some misty Highland dream. Victoria had ushered into being the Highland sporting estate. Some landowners still clung to this way of life even today.

Tabatha opened the door and entered. A thick fug of cigar smoke hung in the air; it was difficult to see the far end of the room. Charles Purdey sat beside a roaring log fire in a huge leather armchair, well into his fifth brandy. Sitting opposite him was a thin, dark-suited man cradling a glass of whisky Tabatha had seen him reluctantly accept from the laird an hour earlier. He was Gerald McCormack, the estate lawyer and a Highland councillor – a man who would argue to the death over four pence.

The lawyer rose to his feet like a spectre rising from its grave. 'What a lovely meal, Tabatha. So kind of you.'

This innocent remark provoked the dachshunds into a furious rage and they hurtled across the room and started to attack his trouser legs with a vengeance.

Purdey rose unsteadily. 'Damn it, damn dogs. I don't know what gets into them.'

He made several enthusiastic but unsuccessful attempts to strike the snarling dogs with a rolled-up newspaper. By now Gerald had spilled his whisky and was struggling to maintain his balance amid the sound of tearing trouser legs.

'Enough!' Tabatha yelled in a voice loud enough to get the attention of most of the stags nailed to the wall in the corridor.

Reluctantly the four dachshunds relinquished their grip on the lawyer's nether garments and slunk off to sit beside the fire. Apologies were given and refused; a replacement whisky was offered and reluctantly accepted. Tabatha grabbed the arm of a heavy leather chaise longue and hurled it in front of the fire with the ease of a weightlifter before throwing herself down on to it with a sigh.

She looked at her husband in expectation, but he was busy

lighting another cigar. 'Just the tiniest G and T, Charles,' she said after a moment.

He rose, fumbling with his matches, and rushed over to the antique drinks cabinet. 'Oh yes, Tabby, quite.'

'I'm not a cat, Charles.' Tabatha lit a cigarette and smiled at the lawyer.

Purdey shuffled over and handed her a drink. 'Yes, of course, my dear.'

Tabatha leaned forward in a familiar way towards the lawyer, who cringed back into his chair, not being a person much inclined to intimacy of any kind. 'Are you a country person, Mr McCormack?' Tabatha took a swig of her gin and didn't wait for his response. 'Of course you aren't. Look at your suit – well, what's left of it. No one of any sense is a country person. We've a house in London, you know. It has heating.'

'Tabatha isn't Highland-bred, you see. Not like us.' Purdey smiled at the lawyer.

McCormack stared at the fireplace, perhaps hoping it would swallow him. 'No, Charles, I'm not. Where I grew up we had heating and you couldn't perform an ice ballet in the bathroom.'

Purdey abandoned his cigar and turned back towards his wife. 'I'm trying to preserve a way of life, my dear. I don't want to be the Purdey who abandoned the estate and let it go to people like bloody Tony Muir.'

'Tony Muir?' McCormack asked.

Tabatha exhaled a huge cloud of smoke in surprise. 'You mean he's not told you?'

McCormack looked mystified, which gave Tabatha the opportunity to launch in. 'His estate is next door to ours. The man's absolutely loaded. He's one of those green people. *Plants trees.*'

Purdey shook his head sadly. 'He doesn't even shoot. What's the point of having a Highland estate if you don't shoot?'

'So he's into forestry – timber, you mean?'

Purdey pointed so vigorously that brandy splashed on to his trousers. 'No, no, that I could understand. He wants to re-forest the place, bring back the native woodland, so there can be beavers and wolves running about. It's ludicrous.'

'Charles, times are changing,' Tabatha said.

But Purdey flushed with indignation. 'He's a bloody lunatic, running an estate in sympathy with the environment. All nonsense – *I'm* the real conservationist. How many red grouse do you think there would be if it weren't for my efforts?'

Tabatha stubbed out her cigarette with venom. She was tired of hearing her husband profess his environmental credentials. 'Until you shoot them, that is.'

'Yes, that's right. What else would you do with them?'

'I suppose he doesn't do any harm?' McCormack said, writhing in his chair. The leather armchair was 200 years old and the springs had seen better days.

Purdey slammed his glass down on to the marble mantel-piece, sending a shower of fine brandy into the fire. 'Harm! Of course he does harm. If his lot ever get into power there's my subsidies up the spout. How do you think I could manage to keep all my lands going if the public didn't pay for them? Answer me that!'

McCormack didn't answer, but Tabatha had to agree with her husband for once. 'You are right there, Charles. People have no idea how expensive privilege is.'

'We're losing everything this country stood for,' Purdey continued without drawing breath. 'And all for some fairy tale of going back to living alongside nature. The modern population can't be sustained alongside wildlife. It stands to reason.'

Tabatha rolled her eyes, sensing that her husband was about to plunge head first into one of his rants.

'Well, I won't let it happen. He wants to bring back beavers – I'm going to prove they spread disease. The dirty little tree-cutters.'

McCormack squirmed again, but there was no escape.

Tabatha took a swig of gin. *Here we go. It'll be the flag first.*

'Do you see that flag?'

Purdey pointed to a tattered ensign hanging limply above the fire. The design on the red flag was faded but it was still possible to discern the figure of a nobleman flogging a peasant.

'My forebears fought under that banner at Culloden,' Purdey said with a gleam in his eye. 'When all others wanted to retreat, the Purdeys stood firm. Our family motto is "ne'er take a backward step". As the redcoats charged and the other clans ran we stood firm. That's courage.'

Tabatha snorted. 'Or pig-headed stubbornness.'

Enraged, Purdey reached beneath the banner and drew an old battered sword. He leaned forward with it as if the redcoats were charging towards him down the great hall at that very moment. 'This is the Purdey sword! Three hundred years of history. Ne'er a backward step!'

McCormack leaned back into his chair, concerned that the furious aristocrat might decapitate him with a careless gesture. 'What happened?'

Purdey lowered the sword; suddenly all the fight had left him. 'Slaughtered to a man, of course. But at least they showed spirit.'

There was a knock on the heavy wood-panelled doors, which once more sent the demon dachshunds snarling across the room.

'Boys!' Tabatha said sternly, and the dogs scuttled back beneath her chair.

'Come,' Lord Purdey yelled.

Donald entered the room. He was carrying a large duffel bag.

'It's Friday, m'lud,' Donald said quietly, keeping his eyes firmly on the polished wooden floor.

Tabatha looked up in alarm. 'Is it? Already? It feels like it was Friday only yesterday.'

'Thank you, Donald, but we don't want to put you to any trouble.' Purdey glanced across to his guest, who was offering a tentative yet benign smile at the old ghillie.

'Och, it's nae trouble, sir. I have the instrument here.' Donald unzipped the holdall and began struggling with a collection of pipes and a tartan bag.

Purdey looked alarmed. Tabatha rose and attempted to shoo Donald away as though he were an errant child. The dachshunds began to emit a low growl.

To all this Donald was oblivious and swung the bagpipes on to his shoulder with obvious delight.

Tabatha was a formidable woman but even she went weak at the knees at the sight of Donald preparing to play. 'Honestly, Donald, there's no need for you to trouble yourself.'

The dogs were now snarling ferociously from beneath the

chaise longue where Tabatha had been smoking moments before.

Donald drew himself up to his full height. 'Och, but ye know the tradition.'

'Oh, what tradition is that?' the lawyer said, brightening.

'Every Friday nicht when Lord Purdey is in residence,' Donald said with a respectful cough, 'the head stalker – that's massehl – must play him a lament after dinner. That has been the practice for over 150 years.'

'Splendid,' McCormack declared.

Tabatha deflated like an oversized balloon. *Victoria started it. Damned woman.*

Purdey slumped back, defeated, and finished his brandy in one huge swallow. 'If you insist, Donald.'

Donald put the chanter to his lips and a sound like a tortured whale escaped. At that instant the four dachshunds whimpered pathetically and bolted shrieking down the corridor. The dogs were the lucky ones; for the humans there was no escape. Donald was unique as a piper for he could summon from the instrument a sound so staggeringly awful that, in another age, his piping would have been used as a weapon of war (today it would be banned by the Geneva Conventions). Purdey drove his fingers into the arm of his chair and writhed.

When at last Donald took the pipes from his lips the instrument subsided like the death throes of some evil beast from the centre of the earth.

Tabatha leapt from her chair with amazing speed, thrust a glass of whisky into Donald's hand, and propelled him out of the dining room while muttering her gratitude to the man.

McCormack mopped his brow with his handkerchief. 'My God, I never heard anything like that.'

Tabatha lit another cigarette and inhaled greedily. 'No one has ever heard anything like that.' She coughed as the smoke caught her lungs. 'Of course, the man went stone deaf a few years ago. Can hardly hear a thing.'

'Must have come as a welcome release,' McCormack said.

SPRING
CHAPTER 3

If you stood on the steps of Inverness town hall with your back against one of the two stone pillars that flank its ornate doors, on a Thursday night at half past eight precisely, you would see Angus Sutherland walking down the High Street to meet with the other members of the Highland Mountaineering Club. Angus had lost count of the number of times he had walked down the hill from his home in the Crown district to cross the river to where the club met in the narrow streets of the town's old fishing community. He was always punctual – something he was proud of. He thrust his hands deeper into the pockets of his jacket, feeling the chill wind blowing up from the River Ness. As he passed the Gellions pub, the town's oldest, a blast of music hit from the open door and the burly doorman nodded to Angus, having grown used to seeing him.

Usually Angus looked forward to his nights with his friends but tonight he felt uneasy as he headed towards the modern bridge across the river. Above him towered Inverness Castle. A Victorian creation, it now housed the courts that passed judgement on the people of the Highlands. Its ruddy sandstone bulk dominated the centre of the town, as if it had been erected as a threat to remind the residents of the Highlands to keep on the right side of the law. The new castle stood on the remains of the old – which Bonnie Prince Charlie had blown up on his way to establish a range of gift shops across the Highlands, and, if the maps are to be believed, visit just about every mountain cave in existence. A tourist bus rolled past as Angus waited to cross to the bridge, its recorded commentary telling a busload of bemused Japanese tourists how a French soldier and his dog had set the explosives under the castle only to be blown across the river when they went off.

As he crossed the bridge Angus remembered when he had left Aberdeen to work in the Highland capital. Thirty years ago, the centre of Inverness was full of pubs and churches. On Sundays the city's residents were to be found in one or the other and frequently both. Now the church congregations were ageing and dwindling, and even the pubs, once packed, were closing down. The town had grown immensely since Angus had moved here. There had been only small one supermarket when he arrived, but the megastores of multinationals now circled the place like invading armies. The heart of the town, like cities up and down the country, was dying, with many old businesses replaced by charity shops.

Normally it took Angus twenty minutes to walk the mile from his house to the bar but tonight he slowed. Something was keeping him back. He stopped, took out his pipe, and leaned against the railings, watching the Ness catching the reflections of the street lights as it rushed by on its short journey from Loch Ness to the sea. His mind replayed the scene of Rory being swallowed by the avalanche. He watched again as Rory reached desperately for the rock and hauled himself to safety. Angus knew that if he'd been a foot further from the rock, or even six inches, the outcome might have been different. They'd been lucky.

Then a memory from almost thirty years ago flooded his mind and he was walking again on a sunlit day towards the savage cliffs of Ben Nevis with a tall, blond young man. James pointed to the ice-wreathed face of the mountain and grinned. Angus found his gaze drawn to a ragged ice gully that swept up the towering cliffs and on until it reached the summit.

He heard James say, 'It looks brilliant.'

Angus was standing at the foot of the climb, hammering a steel peg into a crack in the rock, and then he was watching James climbing, steadily hacking his way up the gully. The day was perfect: cold, sunlit, still. He could feel the rope running through his hands, then he heard James call down from above.

'I'm on belay. Climb when you're ready.'

Angus tore himself away from the memory as if it were a physical thing, slamming a door on a room he dare not enter. He was back standing beside the river, hands gripping the railings so hard they hurt.

He pushed the memory away. 'Stupid old git.'

He should head for the pub, he'd be late, but something kept him rooted to the railings. Thirty years was a long time to wander the mountains. He was slower now, less agile, and when he walked down a steep hill his knees creaked and sent bolts of pain up his legs. Years ago he'd looked at the old men in the mountaineering club and thought them grey, bumbling fools. Now that it was his beard that was flecked, his waist that had thickened, was that what they thought of him? Perhaps that's why fewer and fewer members came on a Thursday evening. When he'd joined the club it was vibrant with young men and women who loved the mountains but now there were few young people like Rory. What did he think of him – an old fool, perhaps? After all, Angus had nearly led him to his death that winter's day. *Maybe I've become an irrelevant relic.*

'Don't do it, Angus!'

A woman's voice cut through his thoughts. He turned and Rory's wife Jen was smiling at him. She was small and slim and her long blond hair blew in the wind as she stood with her hands driven into the pockets of her padded blue jacket.

His face must have shown something of how he felt because she spoke to him again in her lilting Yorkshire accent, gently this time. 'Now then, are you all right, Angus?'

He couldn't tell her he had just been thinking about how close her husband had come to dying at the weekend. He couldn't tell her he was wondering if it was time to put the mountains behind him. He was the president of the mountaineering club, an important position.

He wasn't very good at lying. 'Er, I was just … watching the river.'

He wasn't very good at guessing women's ages either; she was maybe thirty, and her large brown eyes seemed to be able to see right into his soul. 'Tha looks a bit shaken, Angus?' Jen said, reaching out and touching his arm.

Angus wasn't used to being touched by young women and he drew back. 'Nae, I'm fine.' *What would she think of me, an old man standing by the river struggling with memories?* 'Is Rory nae with you?'

Jen must have known he was changing the subject; she had worked long enough as a nurse to know when a man was trying to dip back behind a façade. 'Nah, I've just finished work.

He'll be in't pub already.'

Angus detached himself from the railings and they began to walk the remaining few hundred yards to the little backstreet pub. 'Have you got any further with finding an island to live on?'

Jen shook her head and sighed. 'No, not really. His heart's set on it, though.'

Angus sensed she had doubts. 'Are you nae so sure?'

Jen bowed her head and watched her feet shuffle along the pavement. Angus couldn't see her face and wondered if she was trying to hide something. 'Aye, well, I want to go too, tha knows. But it's a big move.'

'I believe he hates that factory, though.' Angus knew Rory felt caged at work.

Jen stopped, looked at Angus, and nodded slowly. 'Aye, happen he does. I'd 'ave to leave the hospital.' Suddenly she brightened. 'Like my da used to say, if in doubt do nowt.'

Angus wondered if he was prying and saw that Jen was caught between her career and Rory's dreams. He had been so eager to divert the conversation from his own troubles that he hadn't been careful enough.

By now they were amongst the small riverside cottages that had been the homes of fishermen only 150 years ago. The streets were narrow here, and the houses huddled together, as if for warmth. The tiny bar in Celt Street was little more than an old cottage itself, with a triangular extension built on the rear to accommodate the lounge bar.

Jen smiled at Angus as she stepped nimbly up the steps and through the double doors. Perhaps both of them were happier to enter the bar and escape a conversation that was straying into sensitive, personal places.

The lounge bar of the Thistle Inn was comfortable and warm. A deep-green tartan carpet covered the floor and the walls were bedecked with pictures of old Inverness and fierce warriors swathed in plaid. Just like most bars in the Highlands, it had been designed to appear as though Bonnie Prince Charlie had just left, harking back to a mythical time when kilted Highlanders strode about the glens in search of shortbread. It suited the small club of outdoor folk: it was unpretentious, and small enough that on some nights the club would have the place to themselves. Two tables had been pulled close together so people could gather round. The club was a jumble of folk of different ages, many of whom had been members for many years, but few had turned out for the club's weekly meet – Angus spotted only half a dozen or so members in the small bar.

Jen spotted Rory and kissed him affectionately on the cheek. He turned and smiled, and she sat down next to him. 'Made any plans?' she said.

Rory put down his pint. 'Not yet, I was waiting for Angus. He's late.'

Angus came back from the bar with a Guinness for himself, lager for Rory and red wine for Jen. He didn't need to ask; he knew them both so well.

Rory grinned and raised his glass in salute. 'What time do you call this?'

Angus glanced at Jen, hoping she wouldn't say anything. 'It's only ten minutes.'

Jen took a sip of her red wine. 'Nah then, Rory, stop mithering him. He were watchin' ducks. Isn't that right, Angus?'

Angus nodded, looking down into the creamy head of his Guinness, hoping that would be the end of the conversation.

Rory laughed. 'Ducks? Are you becoming interested in wild-life at last, Angus?'

Angus screwed up his face. Perhaps Jen decided it was time to change the subject to save him any more embarrassment, for she turned to Brian, who was sitting at the table in silence as usual. 'Ow do, Brian. Ow's it going?'

Brian nodded and smiled and looked as though he was going to speak for a moment, but then he decided against it and took another sip of his orange juice.

Jen turned back to the others. 'Nah then. What's tha planning for this weekend?'

Rory could never focus on anything for very long and brightened at the question. He was always keen to spend some time in the hills. 'I was thinking of maybe heading to a bothy Saturday night. What do you think, Angus?'

Angus had visited just about every remote shelter in the Highlands. He had spent long, solitary nights in the bothies of Scotland's north coast and shivered in the high bothies of the Cairngorms. His romance with these places began when he was a boy. As a ten-year-old he had followed his father's great rucksack, rattling with pans and provisions, up over the long hill track to a solitary cottage called Shenavall in the heart of the mountains. His young legs had almost given out before they arrived at the bothy door, but once inside he'd felt as if he'd stumbled into a secret world populated by wild mountain men and women.

The impressions of that night still lived with Angus now. He could smell the smoke and almost taste the fried bacon.

At night he'd lain in his sleeping bag, tired but struggling to stave off sleep, listening to the folk in the bothy as they swapped stories of adventures, laughed, drank and sang songs together. As he drifted off to sleep it was as if he were a continent away from the neat semi-detached house he shared with his parents and sister on the edge of Aberdeen.

Ever since that first, tantalising glimpse of another world, he had spent most of his adult life travelling the high places of Scotland, climbing on the cliffs or wandering along the high ridges. His days in the hills had been what gave his life colour; he'd treasured those days, but he realised now that perhaps that love affair was coming to an end.

'I believe I could do with a night in the hills,' he heard himself say, but the words had been spoken from habit. Something had changed.

Rory looked pleased a plan was being formed. 'Sounds good to me. Maybe a hill as well. Do you fancy coming, Jen? You're off this weekend.'

Jen looked a little doubtful. 'Aye, maybe. What's the weather doing? I don't fancy a soaking.'

'Sunshine and showers,' Brian announced to everyone's surprise.

Angus was thoughtful for a moment. 'Well, you know what that means.'

'It could do absolutely anything,' they all said in unison and laughed.

Brian drew himself up. Angus sensed another announcement was coming. 'Better in the west.'

Angus thought for a moment, running through all the

bothies he knew, searching for just the right one. 'Have you been to Bearnais?'

Rory shook his head.

Angus remembered a small stone building set deep in the hills and surrounded by tall mountains. 'It's not too far from Loch Carron. Quite a walk in, you ken.'

'I'm up for it.' Rory turned to Jen. 'What about you?'

Jen hesitated and swirled the red liquid about in her glass. 'Aye, why not? I've not been away for weeks.' Her face brightened.

The decision had been made but Angus felt a strange reluctance he couldn't explain. 'It's a quiet, remote place. We probably won't see another soul.'

Then the talk turned to hills and there was a spirited discussion on which was the most difficult bothy to get to.

***

Rory was halfway through his second pint of lager when he noticed Angus had barely touched his Guinness and was sitting gazing far away. 'You all right?' Rory asked quietly.

Angus nodded and said, 'I'm fine.'

Rory could see he wasn't fine. He was about to press him again when, to his surprise, Angus pulled his pipe out of his pocket and stood up.

'Let's go outside, I need a smoke.'

The air was cold and the wind made Rory pull his jacket tighter as they stepped out into the narrow street. Angus filled his pipe with Irish Cask and they both stood in silence as he went through the ritual of lighting it.

Angus turned to Rory. 'About last weekend … I'm sorry,

that was my fault.'

Rory was surprised by how serious Angus had suddenly become. 'It wasn't your fault.'

Angus shook his head, almost dropping his pipe. 'Nae, it was. I should have known that slope was risky.'

Rory could see Angus was tortured by what happened, but there was no reason for it – they had both known what they were doing when they decided to cross that slope. 'Jesus, Angus, you can't bloody blame yourself for that. I got out—'

Angus cut in with anger in his voice now. 'You widnae have been there if it hadn't been for me. What if you hadnae got out, Rory? What then?'

It was Rory's turn to be vehement now. 'These things happen. You know that. We'd never do anything if we worried about the risk.'

Angus was silent. Rory could tell that he was struggling to contain himself. He reached out and put a hand on the older man's shoulder. 'Angus,' he said quietly, 'don't blame yourself. I knew the risks.'

Angus shook his head slowly, drawing on his pipe. 'I saw you and Jen together and I thought … What would I have said?'

Angus was quiet again and it seemed to Rory that his mind was somewhere else, perhaps a different place or time, but it was somewhere Rory couldn't follow.

Angus smoked in silence for a few moments and when at last he spoke it was in a serious tone Rory had not heard him use before. 'I'm thinking it's time I packed this in, ken. I believe I'm a liability.'

Rory leaned against the pub doorway, watching Angus smoke.

There seemed a finality in his words. 'Angus, don't be … ' Rory tried to find something to say.

Angus tugged at his grey beard and knocked out his pipe on the brickwork of the old inn. He turned and went quickly back into the pub. 'I cannae do that again.'

Rory was left standing in the street wondering *what* Angus couldn't do again.

∗∗∗

Later that night, Jen and Rory sat together on the sofa of their small flat on the edge of the town centre, watching with little interest the end of a panel show. The walls of the flat were covered with mementos of their lives in the hills. Friends smiled out at them from pictures of windy summits. Rock faces glowered down from some images, and in others ice-covered summits glittered in long-remembered sunlit winter days. Here and there Rory's other passions spoke from the walls. There were images of pine martens, badgers, eagles and barn owls alongside the climbing photographs, as if these creatures were more friends from the outdoors.

Jen nuzzled her face into Rory's neck, but Rory was distant and preoccupied. 'Do you think Angus is right – we should think about disbanding the HMC?'

Jen swirled her coffee. 'Well, if there are fewer and fewer members … isn't it kind of inevitable?'

Rory shifted on the sofa as though the thought made him uncomfortable. 'Yeah, but Angus is very proud of the club and being president and everything.'

'Tha have to be realistic. When that club started fewer

people had transport – they had to share lifts. Now everybody drives. There's not the same need for huts neither. Everywhere in't Highlands is accessible in a couple of hours' drive.'

Rory turned away from the TV and gave Jen a gentle squeeze. 'When we go to our island that could be the last straw for the club. Losing us could make up Angus's mind.'

Jen looked back at the TV, not wanting Rory to see the look in her eyes. A group of celebrities she'd never heard of were trying to unscramble the names of film stars on a board. 'Well, that might not be for a while yet.'

'Can't come soon enough though, can it, Jen?'

'No, love, no it can't.'

But moving to some remote island and living off carrots with no TV or internet and having to travel for over a day to get to a decent shop sounded like a nightmare to Jen. She had loved Rory from the first time they had climbed a hill together two summers ago. She liked how earnest he was and how much he cared about the planet. She even loved the fact that he was a hopeless dreamer, and, for a long time had thought that Rory's dream of moving to a remote island was simply a dream. At first she'd told him how great it all sounded, how much she would enjoy the solitude and being far from traffic fumes, not having to fit in with the nine-to-five routine. She had told him that because she thought it was one of his daydreams that would slip away like the idea for an organic brewery and his plan to become a woodcarver. But this particular daydream had taken root in Rory's mind. Now Jen's big worry was how to talk him out of it without it coming between them. As the credits rolled and he said something about bedtime she decided that now was not the time – she'd just have to hope that something brought him to his senses.

She knew his moods now, how he could move from optimism to dark gloom in under an hour. She realised early on in their time together that unless he could counter the dark drudgery of his weekdays in BetterLife with weekends on the hills he would sink into a depression. Sometimes she went with him to the bothies he loved but in the winter she would stay at home and let him have his caveman times out in the wilds. Normally she didn't worry, but his close shave with the avalanche had given her an uneasy sensation – and, sometimes, when her work as a nurse allowed her time to think, she would feel her stomach churn. Those who adventure in remote places can see their fears and know when life hangs in a fleeting moment, but those who wait for them suffer a greater torment, for the imagination knows darker places than reality will ever see.

***

Jen stopped for a moment and adjusted her hood as the rain intensified on the steep climb over the hill to the bothy. They had been walking for a couple of hours, and the forecast sunshine and showers had turned out to be a five-minute sunburst followed by two hours of heavy rain. Now water was beginning to find its way past the defences of her cagoule, seeping down her neck and wriggling like cold fingertips up the sleeves of her jacket.

Rory, as always, was striding on ahead. Looking down the hill Jen could see Angus plodding steadily upwards a few hundred yards behind, obviously struggling. She stopped to let him catch up, not wanting the older man to feel he couldn't match her pace.

Angus came to a halt beside her. He was red-faced and sweating heavily underneath his waterproofs. 'Bloody hell, I'd forgotten it was this steep.'

Jen laughed. 'Aye, it's always steeper than tha thinks.'

The rain eased a little as they neared the crest of the ridge, although the wind increased and great sweeping curtains of rain swept across the sky. Rory was still ahead when he stopped, yelled something Jen couldn't hear, and pointed to the sky. Jen pushed the hood of her cagoule back and gazed upwards. A hundred feet above them a dark shape weaved out of the clouds.

'Eagle!' Jen yelled.

The huge bird soared out of the sky, so close now they could see the feathers on the tips of its wings spread like the fingers of a hand. Jen had seen golden eagles before but every sighting was special. For her no other bird symbolised the wild places so much as the eagle, with its majestic size and effortless mastery of the air. Every day she saw an eagle was a memorable day; to glimpse these wild creatures was a privilege.

She'd been with Rory when she'd seen her first eagle. Two years ago now – the first time they had gone out into the mountains together. She had thought him a little odd and awkward at first. He seemed never to settle in a chair but then she had seen him moving in the mountains and noticed he had a remarkable grace. Out in the hills he was at ease; up here she saw a different Rory, full of passion for the outdoors.

The eagle that day had taken her by surprise. She remembered him turning to her, eyes bright with excitement. 'Look, look there,' he'd said. She had tried to follow his gaze but could see nothing. He'd laughed at her, told her she was blind. Then he had turned her gently towards the eagle, and stood behind her so that she could look along his arm to see where he was pointing. Then she'd seen it. It was like catching sight of God.

She had felt him close to her as they watched that bird. She could feel the excitement coursing through his hands as he too marvelled at the creature.

'Look up, Jen. You need to learn to look up.'

He was right, she'd known that; she had spent most of her life looking down. She'd been fastidious in school, worked hard at university, and when she'd started nursing she was always the one with her nose in books working for better grades. She had never been one of those girls coming back from a club late at night, laughing too loudly.

Rory had taught her to look up and she had loved him from that moment.

Back in the present, they watched the eagle until it vanished over a ridge, covering in only a few minutes a distance it would take them hours to hike. They walked on and at last reached the top of the ridge. Jen had never been to this part of the Highlands before and now the landscape opened up before her in an unending sea of rolling ridges, deep glens and snow-flecked summits.

Angus pointed. 'There's the bothy doon there.'

Far below, Jen could make out the grey slate roof of a building nestling in the valley floor.

Angus grabbed his map and peered at the scene below. 'That track was nae here last time.'

Rory pointed to another track. 'That looks new as well.'

'Bloody hell, I can see at least four new tracks.'

An hour later Jen felt the soles of her boots hit the hard gravel of the first track. From above, the roads had looked like thin grey lines, but close up they were wide enough for two lorries to pass and their surfaces were like compacted concrete.

'This place used to be quiet and peaceful,' Angus said in a low voice. 'Now look at it.'

The tracks were torn into the flesh of the earth. Great ditches either side showed black where the peat beneath the heather had been savaged by the steel teeth of a bulldozer. The streams that would have crossed the track were corralled into culverts or diverted away. What had been open hillside was now shredded by long, man-made scars.

Rory shook his head sadly. 'Look at the place. Devastated. Like I told you, these bloody hill tracks are everywhere.'

Rory led off the mile to the bothy and Jen followed with Angus close behind.

They had only walked a few hundred yards when something roared behind them – a massive, mechanical sound – and the ground shook. Jen turned. A digger the height of a double-decker bus was bearing down on them. The beast ground to a halt and stood with its diesel engine rumbling. The window of the cab about ten feet above them slid open and a cheerful-looking driver popped his head out.

'All right. Where are you geezers going?' The driver sounded like he had come a long way from south London.

'We're heading for the bothy,' Rory yelled up over the sound of the machine.

A bemused look came over the driver's face. 'Blimey, bit of a rough gaff that. You wouldn't catch me stopping there.'

Angus called up to the driver. 'What are all these new tracks for?'

The driver beamed, as though delivering good news. 'For the hydro schemes, mate. There's four of 'em 'ere.'

'Used to be such a quiet place, this.'

'Will you bury the tracks when the work's finished?' Rory asked.

The driver looked puzzled. 'Nah, mate, that's not on the cards. Why would you want to do that?'

It seemed so obvious to Jen. 'To restore it to the way it were, like.'

The driver scratched his head and surveyed the hillside. 'There weren't nothing here before we put the tracks in, darlin'. Place was just bleedin' empty.'

The driver was right. Before the hydroelectric schemes, the glen was empty and silent. Save for the sheep, the deer and the occasional walker, there was nothing. We have yet to learn that nothing is something.

At last they found the stone shelter nestling beside a small stream. It was a simple one-roomed building with a couple of crude benches and a table at one end. Jen was happy to sit on one of the benches and watch as the two men went into their well-rehearsed routine, unpacking stoves and pans and spreading out sleeping bags. She smiled to herself as she watched Angus setting up his little stove. Laura, Angus's wife, had told Jen that she wouldn't let him into the kitchen at home, yet out here he took a delight in sorting out his food and preparing a basic meal.

Laura never came to bothies – she had once described them as dirty little hovels – and couldn't understand how Jen could like them. Jen's hospital shifts meant that weekend leave was infrequent so she couldn't join Rory as much as she'd have liked. But she did enjoy bothies, even if they were rough, masculine places full of dust and spiders. Besides, Rory was always happy in a bothy, and she liked to watch him forget about life in the factory, even if only for a few hours.

At last darkness fell and the three of them relaxed in front of the small fire as the world shrank to the arch of their candlelight.

Angus blew out a circle of pipe smoke. 'I see they've finally stopped.'

Rory peered out of the window and into the darkness of the glen, searching for headlights. 'Yeah, they must have packed it up for the night.'

Jen looked at her watch. 'Six thirty. They'll be off home for their snap.'

Angus turned to her, looking quizzical. 'Fit?'

'Snap? It's Yorkshire for a meal,' Jen said with a laugh.

Rory was also laughing. 'One broad Yorkshire and the other an Aberdonian. It's a miracle either of you can understand the other.'

Angus shook his head. 'I'm nae broad at all, you ken. I've lost most of my Aberdonian. My faether spoke the Doric – yer two from south of the border widnae have understood him.'

There was a warmth in the way Angus spoke. He had no prejudice against them because of their English roots, unlike some Jen had encountered in her work at the hospital who saw nothing wrong with labelling her a white settler.

Angus put down his pipe. 'This place used to be so peaceful. My faether brought me here when I was nowt but a loon. Now there's great bulldozed tracks all over it.'

'I see them everywhere I go,' Rory said. 'New tracks tearing up the hills.'

Jen took a sip of wine and felt the rich liquid bring a warm glow to her stomach. 'Aye, but these are for hydroelectric schemes. Maybe we has to accept 'em if we want green energy.'

Rory shook his head slowly. 'No, no way. The amount of energy these hydro schemes generate is tiny. Not worth the damage they do. They're everywhere you go.' He turned to Jen. 'You remember that remote bothy we went to last year when we saw the seal pups in Sutherland?'

She remembered a long boggy walk and rain beating on the windows out of night so dark it seemed able to swallow light. Then, in the morning, they had walked to a small shingle beach and crouched in the heather watching tiny helpless seal pups being tended by their mothers. Rory had been so excited. He had grown up on Merseyside, beside a river so polluted it was slick with oil. Seeing somewhere so wild had made him like a small boy again and she had laughed at him.

Jen nodded.

'Well they're building a bloody space-launch site there. It's a disgrace.' Rory stood, as if the thought of it made him restless.

Angus scratched his beard, something Jen realised he often did when troubled. He took a puff on his pipe while he formed his words. 'It's nae that simple, you ken. Sutherland's population has been in decline for a long time. The whole area is struggling to survive. It'll bring jobs, revive the area. I don't want tae see the place spoiled, same as you min, but what can yer do?'

There was silence for a few moments until Angus spoke again. 'Do you ken how few folk are coming tae the club these days?'

'Aye, 'appen it's been quiet these past weeks,' Jen said.

'When I started climbing,' Angus said, 'you went to a club if yer wanted to find the whereaboots of a bothy or a climb. Yer asked folk. Now it's all online. Folk don't meet each other any more.'

Rory shovelled a few lumps of coal on to the fire, which hissed gratefully back. 'People still meet in bothies, places like this. Maybe this is the last bastion of the real world.'

Jen laughed. Perhaps Rory was right; perhaps these crude shelters did give people refuge from a world full of noise and technology.

'It'll pick up in the summer, you'll see,' Rory said.

Angus suddenly became grave. 'Well, if it disnae I am seriously thinking we'll have to disband the HMC.'

'Oh, come on. It's not at that stage yet,' Rory said.

There was something different about Angus, Jen thought. In the space of only a week he'd aged, grown less sure of himself. This was a new Angus. The old Angus had always been confident and determined but somehow he had lost the old drive.

Rory got to his feet. 'You need some tea, man. Cheer you up.' He picked up the kettle and stepped out of the low wooden door to fetch water.

Jen watched Angus staring into the fire. 'He really respects you, tha knows.'

Angus looked surprised. 'Really?'

'Aye, says he knows nowt about the hills compared to thee.'

Angus smiled and made an odd embarrassed grunt. Jen was pleased to see he'd brightened a little.

The door opened and Rory came back into the sanctuary of the bothy sloshing water from the kettle. 'There's someone coming, I could see a head torch.'

Jen lit the stove and put the kettle down on to the hissing flames.

There was a rattle and the door burst open. A young man, dressed in brand-new lightweight kit, entered the bothy backwards, filming himself with his phone. 'Here I am, day twenty-three, entering Bearnais bothy,' he muttered into the device.

He turned the camera on the three bothy dwellers sitting by the fire. 'Here are some people staying in the bothy.'

There was a beep from a device clipped to his waist, and then an electronic voice spoke. 'You have stopped. You have stopped.'

The young man pulled down his hood and grinned.

Angus whispered to Rory, 'Bloody hell, he looks about twelve.'

Jen called over to the young man. 'Nah then, Luke Skywalker, does tha want a brew?'

***

Monday morning arrived like an unwelcome guest and Rory's weekend visit to the bothy retreated into memory. The factory corridor was long and the polished floor echoed with the footsteps of incoming workers. Rory followed the long line of men and women. Most would be heading for the mind-numbing tedium of the production line, others for only marginally less mundane jobs in offices with doors marked Engineering or Product Development. It didn't really matter where you were going on this production line of humanity; whatever you were doing you were a cog in the machine of this vast American multinational. BetterLife was the biggest commercial employer in the Highlands, and the only real option if you weren't in the council or the NHS or didn't want to spend your life making endless hotel beds.

Rory slammed a coin into the coffee machine and hit the button for black. There was a mechanical gurgling and a dark liquid poured into his cup with the plastic aroma of chemically simulated coffee. He popped the top on to his cup, sighed, and tramped off down the corridor. Yesterday he had been a free man; today he was a worker bee heading into the hive. The latest memo had said that you had to walk on the right-hand side of every corridor, just like they drive on the right in America. This, the memo had said, was for 'staff safety' and to 'ease the passage of personnel in the production unit'. Rory resented being told how to walk down a corridor by corporate America.

In a few years' time this corridor would probably be full of robots beeping along in orderly lines, and folk like him would be out of a job. Until then, however, the corporation did its best to turn its human workers into machines.

At last he came to the small, windowless laboratory where he would spend the next eight hours. Inside, the hum of the air conditioning greeted him and he reluctantly wedged himself into the plastic chair. There was nothing natural here. Everything was hard metal or plastic. He had learned a few tricks over his two years in this place that got him through the day. Trick number one was to look at the clock as little as possible. Trick number two, he put on headphones and vanished into a podcast. It didn't matter what he listened to; the podcasts opened up an unregulated world into which his mind could retreat and he could explore at will. Most of all he listened to stand-up comedy. Even while his fingers were typing up the results of thousands and thousands of tests his mind could be in a cellar in New York listening to a comedian undermining the very system that Rory hated. It was his little secret world, and when he couldn't escape to the hills he would go there in an attempt to preserve some sanity.

The morning slipped past him in a glacial tide of figures until at last lunchtime approached and he could head down to the canteen. The company had introduced a vegan option on the menu after he had badgered them for months, but he didn't trust them to have avoided all animal products so he always brought his own lunch. Today Jen had made him a couscous salad. She was a vegetarian, but he could never persuade her to take the final step and turn vegan – she liked cheese too much for that. He ate with one hand and opened the browser on his phone with the other.

There was news on Twitter about legislation heading for the Scottish parliament that would mean landowners would have to get planning permission for hill tracks. Rory had been following this for months. At least that would be a start in controlling the unregulated growth of hill tracks across the mountains of Scotland – better than the current position, where any hill track was acceptable provided there was an agricultural reason for it (which was often dubious at best).

It saddened him that the tracks they had seen at the weekend must have already had planning permission. Maybe the new legislation would be a beginning.

The memory of the hill tracks slashing across the landscape, and the rumble of the earth movers ripping up the hillside, flooded into Rory's memory. He stood in the treeless glen again and saw the barren hillsides rolling endlessly away, a wet, green desert. The mountains bereft of life, empty of creatures save for countless sheep and deer, their teeth nibbling any attempt by the forest to return. This was the earth scoured, lifeless, near dead. *What could this place be like?* he wondered. *What if the trees came back?* Then he stood in that glen again but saw a different place. This time there were trees reaching up high above him like a rolling sea, their twisted boughs hung with lichen, birds flitting to and fro as they sought out seeds and insects. On the forest floor roamed wolves and lynx while the air itself was thick with insects and the calls of birds and animals. *It could be like that*, he thought, *not everywhere, but at least somewhere. We have sacrificed this land to the gods of greed. We should ask these mountains for forgiveness and give them back their children.*

Rory was lost in thought when he became aware of someone sitting opposite him at the table. He looked up and saw a thickset, middle-aged man. His skin was slightly tanned and he wore a beige suit. Rory realised at once that he was one of the managers BetterLife had brought over from America. They always wore beige, as though any other colour might make them stand out as an individual. When Rory started at the factory two years ago, he had tried to talk to these people, but with few exceptions he found them to be corporate clones. Most, from what he could tell, had little life outside of work and no family connections; they were married to the company. Rory had labelled them 'the beige people'.

The American made eye contact. 'Hi,' Rory said.

The American hesitated eating his steak. 'Hi.'

Rory could see he was struggling to find something to say. Eventually the American spoke again. 'Did you have a good weekend?'

Rory realised that the American hadn't the slightest interest in what he'd done over the last couple of days. Perhaps he'd read the question in some company manual about how to talk to underlings.

'I went to a bothy.'

The American chewed his steak. 'Bothy?'

'They're basic shelters in the mountains. No power or running water, just a house with a fireplace. Old shepherds' cottages,' Rory said between mouthfuls of couscous.

The American looked at him in amazement. 'And you sleep there? Don't the old shepherds mind?'

Now it was Rory's turn to grin. 'No, I mean they used to be shepherds' cottages, years ago. But not any more.'

'Oh right. So what does it cost?'

Rory shrugged; it felt like he was passing on a secret. 'They're free.'

The man in beige struggled with that idea even more. Things are never free in corporate America. 'So you just book, right?'

'No, you just show up.'

'What if there's someone else there?'

'Sometimes there is, sometimes there isn't. You just kind of share.'

'So anyone could just show up. Sounds a bit like the Wild West.'

Rory cast his mind back over a hundred bothy nights and nodded. 'Yes, sometimes it is.'

They lapsed into silence and Rory returned to browsing Twitter. There were lots of photos from mountain people he knew, mostly of folk grinning beside bothy fires or waving in spectacular snow scenes. Mountaineers are fascinated by snow's ability to transform landscapes. Then he read a comment about a west-coast farmer who had spoken at a conference, warning that sea eagles would migrate to the east coast of Scotland and bring catastrophe to the farming industry.

Rory knew there was no evidence for this, but the landowners' lobby was powerful and would make claims without the slightest scientific support.

He couldn't resist tweeting back: 'It sounds like pretty much anything could bring an end to Scottish farming. Sea eagles could do it, beavers, a rise in the butterfly population or a vegan cycling past.'

Then he got up from the table and headed off for another four hours in his work cell.

CHAPTER 4

The glen was silent that morning, filled with the dull grey light of winter, as Donald steered the Land Rover up the hill on to the higher slopes of the Black Moor. The hard gravel of the track jolted the steering wheel painfully in his hands, but, despite the discomfort, he was glad of these hill roads.

Donald looked over to the passenger side where Hamish sat awkwardly, his big frame wedged into the seat, typing a message into his phone.

'I'm awful glad we dinnae have to walk all the way up here. I'm glad we put these tracks in, Hamish boy.'

Hamish didn't look up, intent on his phone, playing with his ginger locks with his free hand. 'Aye, right enough.'

'Whit? Speak up.' Donald watched Hamish typing. 'Youse always on that bloody phone.'

Hamish shrugged. 'Donald, will you get off ma case? Everyone my age uses the phone all the time. Have you even got a phone?'

Donald fumbled in his jacket and produced an old phone. 'Of course I have. Who you messaging anyway?'

Hamish grinned mischievously. 'I'm just telling some lassie how gorgeous she is, and letting me mum know what time I'll be home for tea.'

'This yer bird, then?' The phone slipped from Donald's

fingers and fell into the footwell. 'Bugger.'

Hamish laughed. 'No, just some girl I met in Thurso when I was on the deerstalking course last year. They fair go for the tweed, you know.' Hamish stroked his jacket affectionally.

Donald shook his head. Two weeks in a classroom and the young man thought he knew everything. How could he make him understand things that took a lifetime to learn? Hamish was young and ambitious and, just as young men always do, he wanted everything and he wanted it now and wasn't too worried about how he got it. It had taken Donald a long time to work his way up to head stalker, many years of patiently learning his craft. It was thirty years now since he'd taken the train north from his home in Glasgow and made a life here among these hills. That's a long time out in the wind and rain. He'd been back to Glasgow a few times but felt like an outsider there amongst all the clamour and the people. The place of his birth held little for him now, save memories of his father staggering about drunk and swearing at him, or of his mother, pale, sad-eyed and slowly turning her back on the family.

They reached the end of the track. Donald brought the vehicle to a halt and turned off the engine. Excited barking exploded from the back as the three spaniels riding in the cage, realising that their time to roam the hills was at hand, bounded about with excitement.

'Quiet!' Hamish yelled.

Donald stepped out of the Land Rover. The cold March wind took him by surprise and his fingers quickly numbed as he shrank into the comfort of his tweed jacket. Great grey clouds rolled above them, and white waves of snow wandered across the mountains, dusting the tops as they passed.

The dogs leapt out on to the track, their noses filled with a thousand scents and tails wagging with uncontainable excitement. Hamish handed Donald a shotgun and carefully examined the weapon he held in his own hands, heavy and cold and smelling of oil. The stock was engraved with hunting scenes and thistles.

Hamish turned the gun over in his hands. 'Have you ever wondered why Scotland has the thistle as an emblem? I mean, the English have got the rose, the Welsh have got pretty daffodils—'

'Aye, an' we've a flower that can stab yer in the arse,' Donald cut in. 'Says a lot aboot Scotland, that.'

Donald cast his eyes across the hillside. Where they had parked, the yellow grass, flattened by the snows of winter and starved of life by cold and darkness, was clear of snow. Above them the browns and yellows of the heather and grass gradually gave way to white. At first the snow lay in patches here and there, filling hollows and crevices that sheltered it from the sun. Higher, the air became frigid, and the patches of snow began to join together until the ground had turned white and the northern land was locked in winter.

'Let's not go too high. We'll never see the buggers in the snow,' Hamish called to the older man, speaking loudly and only when Donald was looking his way. The young man had learned to cope with his companion's disability, although Donald noticed that he avoided his bagpiping as if it were a curse.

Donald rubbed his hands against the cold, watching as Hamish set off up the hillside without waiting for him, fit and keen and impatient. They kept a couple of hundred feet between them as they traversed the slope, close enough so that nothing could escape them but far enough away that an errant shot would fall harmlessly into the heather. They moved together across the hillside as they had done countless times before, dogs and men moving as one unit. Occasionally one of them would call a dog to turn or bid it stop and wait for the group. Mostly there was no need. At times a telepathy existed between men and dogs, a relationship born of thousands of years of hunting together.

'Three o'clock!' Hamish yelled.

Donald turned and saw the flash of white as a mountain hare – hiding until the men and dogs had come too close – broke cover and bolted across the open moor. Donald spun as the hare rose, bringing the gun to his shoulder and firing in one movement. The impact of the shot lifted the hare skyward before it fell to the ground and lay writhing for a few seconds before death claimed it. Donald felt no emotion whatsoever as the animal died. He took no pleasure in killing, but it had to be done.

'Nice shot,' Hamish yelled over the wind. 'You might not be able to hear but by God you can still see.'

The two men laughed, and Donald picked up the lifeless creature and put it in his bag before they moved on. They carried on like that for an hour or so, taking a hare every few minutes. Their pace was slow and deliberate, for this kind of killing is not something to be rushed. High on the hills, when the mountains are thick with snow, the white fur of the mountain hare allows it to vanish like a ghost; but low down, where the grass lies bare or when the winter snow is only fleeting, it is easy for both man and eagle to spot against the purples and browns of the heather.

After an hour they came to a flat area where the tussock grass and dead heather of the winter gave way to short green grass. The remains of low walls, like the skeletons of whales stranded long ago on a beach, showed where houses had once been. Some gable ends remained standing like stubborn old men, memories of chimneys and hearths. In other places the taller structures had long since given way to the weather and lay as fallen corpses.

Hamish cast his eyes across the ruins. 'What do you think used to be here?'

'Houses, yer nugget. People used to live here.'

Hamish was looking at his phone when he spoke. 'In olden times, like? I wouldn't fancy it. Must have been a rough life up here.'

Donald lit one of his acrid liquorice roll-ups and let the smoke filter down into his lungs. 'Yer think so? I was brought up in the Gorbals. I'd have taken my chances up here given the choice. We was always broke an' ma da was on the swally.'

But Hamish, leaning against one of the ruined gable ends, was intent on his phone. 'No signal again.' He glanced up and looked at the fallen walls. 'Who'd want to live up here in the hills?'

Donald spat out a mouthful of smoke. 'Yer eejit. Do yer not know people used to live all over these hills? Look in any glen and yer'll find ruins like these.'

Donald looked at the ruins and they stirred a distant echo of a memory. Once there had been families living here, a small community of folk who made their living from the land. A poor people but not an ignorant people, whose children were educated in schools dotted in the villages below. Folk who loved these places.

Hamish stood on one of the walls and held his phone aloft. 'Ah, one bar! Why did all these people go, then?' His phone made several pings and he peered at it intently.

Donald couldn't believe the young man's level of ignorance and was beginning to lose patience with him. 'Have you nae heard of the Clearances? The landowners shifted folk off the land to raise sheep. Scattered them across the world, Canada and the like.'

Recognition flickered across Hamish's face. 'Oh, we did something about that in history right enough.' He jumped down off the wall. 'I wasn't into history. All those bloody battles and dates.'

Donald tossed his cigarette into the grass. 'Aye, we better get moving again.'

The dogs tumbled over the walls and licked at his feet, anxious to be running again. As they fanned out on the Black Moor, it occurred to Donald that here he was, clearing the hills of hares for a rich man just as people were cleared from this place hundreds of years ago. In all that time not much had changed.

Folk forgot but perhaps stones remembered. Somewhere in this jumble of stones that were once people's lives, Donald wondered, did a memory remain of a day when the air filled with smoke and weeping? A day when lives were uprooted and the glen fell silent as the empty houses crumbled?

They walked on in silence, seeking out the hares and killing as many as they could. A few evaded the shots and bounded away across the moorland to the safety of the summit. Most, disturbed by the dogs, were quickly dispatched by the guns.

At length they huddled together in a shooting butt for a break from the wind. Donald fumbled in his bag and produced some biscuits, which he offered to Hamish, but the younger man silently waved them away and busied himself with his phone again.

'No signal. You know the problem with this moor? It's too small. I wish we had the forested section to the east of the river. That bit of Muir's land where the bothy is.'

Donald looked at Hamish, surprised. 'Do yer nae know the story?'

For once Hamish gave Donald his full attention. 'What story?'

Donald settled himself comfortably against the wall of the shooting butt, knowing the tale he was about to tell might take a little time. 'Well now, haud on, it wis over forty years ago. When Lord Purdey was nae much older than youse ... '

***

The fire in the great hall roared and spat as Lord Purdey stared at his hand of cards. He glanced up at Tony Muir, who was calmly folding his deck into his pocket. Muir had long hair and sported a goatee beard and moustache. His jacket had broad lapels and he wore it over a flowered shirt with flared jeans. Purdey thought he looked incongruous in the high-backed leather chair, like a prog-rock star who'd hit it rich with his first album and splashed out on a stately home he had no idea what to do with.

Purdey wore a kilt and was grateful for his heavy tweed jacket when the wind rattled the dark windows. He looked down at his cards for the third time, pleased to see such a high suit.

He smiled across at Muir with all the evil intent of a crocodile. 'Your call.'

Muir leaned back in his chair, deep in thought, and hung one leg over the arm. 'Yeah, I'm giving it some thought, man.'

Inwardly Purdey grinned. Muir had no class at all. This son of a biscuit maker – a *baker*, in fact – was new money, and he wouldn't know class if it hit him in the arse. It was only a card game but Purdey would enjoy humiliating the little hippy. He looked down on the pile of notes on the table (which must, by now, amount to several thousand pounds) and imagined it snuggling warmly in his jacket pocket.

Muir flicked back his hair and threw another wad of money on the table. 'OK, I'll see you.'

Purdey smiled gently, like a cat might smile at a mouse, and slowly placed his cards on the table for Muir to see.

Muir's eyes widened. 'OK, you got me, take it away.' He looked frustrated now and he got up as if to head home.

'Not *leaving* already, are you?'

The young businessman shrugged. 'It's late, man.'

Purdey eyed him carefully, noted how tired he looked, a bit drunk perhaps. This would be the perfect time to teach him a lesson.

'Come now, just another hand before bed.'

Purdey poured another brandy, and Muir shrugged.

'All right, but let's put some sounds on.' He walked over to the stereo system and moments later Pink Floyd's *Dark Side of the Moon* echoed through the large wood-panelled drawing room as he sat down cross-legged and began rolling a joint on the album's cover. 'Can't get enough of this band.'

Purdey dealt out the cards and picked up his hand. *Saints preserve us*, he thought. *Another great hand.* They began betting again. As Purdey looked on, Muir drifted away on a cannabis cloud, carelessly adding his stakes to the pile.

With the money piling up, Purdey noticed in alarm that even the cash that he had won was running low. The baker's son seemed to have an endless supply of folding money in the pockets of his absurd denim jacket and appeared only too happy to ladle it out. Purdey glanced at his hand; it was good, not perfect but pretty good. He looked at Muir, but his eyes were glazed – probably on some psychedelic trip. He was about to be taught a lesson. You can't play poker from the far side of the moon.

Muir's vacant expression turned to a slow, knowing smile.

'You're bluffing, Muir – don't think I don't know it.' Purdey's voice echoed around the hall.

Purdey took another sip of brandy and slammed the glass down on to the table. At the sound the deerhound at his feet stirred and opened one eye, as if fearful that the house was under attack; reassured, it drifted back to sleep.

Muir smiled at Purdey through the gathering cloud of dope smoke. 'All right, I'll raise you 500 pounds.' He took another blast of the joint and coughed. 'Don't worry, man, it's just a game. If you don't have the cash on hand to take me on tonight, we can just forget about it.'

Purdey grunted, the ends of his moustache twitching.

He'd be damned if he would be outdone by some fool who looked like Dylan from *The Magic Roundabout*.

'A Purdey never takes a backward step. I tell you what I'll do. I'll match your 500 pounds with land. I'll see you, and if I lose, all the land to the east of the river is yours. Is that fair?'

Now it was Muir's turn to look uneasy. 'I can't expect you to do that.'

Purdey scowled at Muir sitting across from him in his hall, all flower power and cockeyed confidence. 'I think you'll find I know just fine what I can and can't do.' He leaned forward across the table, his eyes bulging, prodding at the baker's son with his finger. 'All the land to the east of the river, between both our estates. See you!' With that he slammed his fist down on the table with such force that the deerhound rose from the floor and walked away on stiff legs.

Muir sighed. 'Well, if you insist.'

'Oh, I do insist, I most certainly do.'

Muir casually laid out his cards face up on the table. They were all spades and all five were in sequence. 'A straight flush. What I'm wondering is, can you top that with a royal flush?'

Purdey's face darkened. He could not have looked grimmer if he had been standing on the battlefield at Culloden and had spotted a redcoat with a Kalashnikov. He rose and hurled his cards into the fire.

Muir watched the cards burn. 'Now we'll never know. I won't hold you to—'

Purdey cut in, sensing an insult. 'I'll not let it be said that a Purdey doesn't keep his word.'

With that he lifted a napkin from the table and began to write.

That evening Muir walked unsteadily out into the cold Highland dawn with 2,000 acres of the Purdey estate crumpled in his back pocket.

***

For once Hamish had no thought for the text messages from a girl from Thurso on his phone. 'Bloody hell, he lost the place on a card game?'

Donald nodded. The wind was picking up now and he zipped up his coat. 'Aye, but if youse fancy keepin' yer job I widnae remind him of it.'

The young man nodded and put his phone back into his jacket pocket. Suddenly he looked up at a flash of movement. 'Bloody harrier.'

Donald produced a small pair of binoculars from his pocket. The white wing of a bird fluttered for a moment above a burn. 'I see him. Think it's the same male we saw last week.'

Hamish picked up his shotgun. The bird was well out of range but Donald saw that he might have a chance if it came closer. Only a few feet above the grass, the harrier hugged the ground in its flight, rising and falling with the contours beneath its black wing tips showing occasionally as the bird hunted for voles and grouse. The two ghillies followed the bird's movements as it came closer. Just as it appeared to be about to fly over them, the creature veered away and climbed high with beats of its powerful wings. The bird effortlessly covered hundreds of yards of moorland, moving with grace and ease through its domain. Soon it was a dot in the distance and then it was gone.

Hamish put down his gun. 'He saw us.'

'Aye, they're canny birds.' Donald lit another cigarette and let the smoke escape from his mouth until it was taken by the wind.

Hamish stared into the distance. 'Numbers are down.'

'Aye,' Donald sighed, 'and that bird'll be part of the reason.'

'Purdey won't be happy.'

'Maybe the breeding season will be good this year.' Donald hurled his cigarette stub into the heather.

Hamish picked up his bag and gun. 'No bonus for you and me if there aren't enough surplus grouse for the guns.'

'We are doing OK keeping the hares doon. At least that should mean the grouse are healthy. Keep doon the risk of tick infection.'

'That's all very well, but vermin like that damned harrier will hit us hard. That bloody bird'll not cost me money.'

The young man opened his jacket and took a white plastic bag from the poacher's pocket, which he passed to Donald.

Donald eyed the bag suspiciously. Peering inside, he found four dead voles – sad, limp things. 'Haud on. Have yer baited these?'

'There's enough in any one of those to take out that bird, no bother.'

Donald looked up, eyes wide. 'Are you a header? You could get the jail for these.'

Hamish snatched the bag back. 'Oh come on, who's going to see us out here? Besides, what are we supposed to do? Purdey wants his grouse numbers keeping up, and getting rid of a couple of those harriers could make all the difference.'

Hamish got up from the butt and walked quickly out on to

the moor, pursued by Donald.

'Hamish, I'm having nothing to do wi' this.'

Hamish turned, his fist clenched. 'Everybody does it. How the hell are we supposed to do our jobs without getting rid of predators?'

Donald was about to reply when something glinted on the hillside a few hundred yards away. All his years on the hills had given him a sixth sense for anything moving and he knew instantly they were being watched.

Donald froze. 'Hamish, there's someone on the hill.'

The big keeper stopped in his tracks and hid the bag of voles in the heather. 'Who is it?'

Donald was peering through his binoculars by now and he could see a man lying in the grass; moments later he recognised him. 'Jesus. It's that RSPB bastard.'

Both men turned and strode to the Land Rover. Ten minutes later they were standing in Lord Purdey's office.

**

Purdey listened to what they had to say, and then rose from his chair in a rage.

'What, the RSPB on my land again?'

Donald nodded. 'Aye, sir.'

'But I told them everything we do here is above board.' Purdey paced his office. 'Right, I'm going to have it out with them. With me, men.'

Purdey stormed downstairs and made his way to the garage, his keepers struggling to match his determined pace. He slammed his Daimler into gear, floored the accelerator, and hurtled out of the driveway with the two startled keepers in the passenger seats. He headed up the road and moments later spotted a small white van headed down the hill towards him with the blue logo of the RSPB on its side. Purdey slammed on the brakes and leapt from the car.

The occupant of the invading vehicle, a man in his thirties in a green fleece, lowered the window. Purdey thought he looked like a prize-winning prat; the very sight of the man sent his blood pressure through the roof.

Purdey wagged his finger. 'Look here, have you been spying on my men?'

The young man smiled back, undeterred. 'Not spying, sir, just monitoring activities.'

'But I explained to you chaps a few weeks ago that nothing untoward occurs on my estate.' Purdey tried to sound authoritative, but was nonplussed to see that it had little effect on the cheerful little RSPB fellow.

There was another diplomatic smile. 'Well, you see it's our policy to monitor estates.'

'How dare you monitor *my* estate? I've told you nothing is wrong and that should be an end to it. I won't stand for this, I can assure you.'

With that Purdey turned on his heels and marched back to the car. The young RSPB man waved and drove off.

Purdey watched him go, also giving a hand gesture (although his wasn't a wave). Donald and Hamish looked at each other. There was silence in the vehicle for a few moments as Purdey chose his moment.

'I don't suppose that if that RSPB chap were to go on to the

Black Moor he'd find anything ... ' Purdey paused, choosing his words carefully ' ... *untoward*, would he?'

Donald and Hamish glanced at each other again; it was Donald who spoke. 'Nae, nothing at all.'

A silence hung in the air like a malevolent fart; no one could see it but everyone knew it was there. Purdey waited. He knew that Donald had answered too readily and sensed that there was something else to be said.

Hamish fidgeted with his phone. 'Well, nothing except maybe a few voles.'

Purdey turned in the driver's seat, puzzled for a moment. He so rarely had to dirty his hands with the *details* of his estate. 'What? The bloody moor is teeming with voles.'

Hamish coughed, and appeared to have difficulty forming his words. 'Yes, sir ... but these are ... poisoned voles. For the vermin, you see, harriers and the like.'

Donald spoke up now. 'They're taking a toll on yer grouse, sir.'

Purdey thought for a moment, wondering how best to tackle the problem. Then he smiled to himself as the solution coalesced in his mind. 'I have it! Attack is the best defence.'

***

The afternoon was cold and a fine rain filled the air like goose down. Simon Partington sat in his car and let out a long, disappointed sigh as he looked out at the rain coursing down the car windows. He was a large, soft-faced man with long dark hair. When he entered journalism he had seen himself as arts correspondent for the *Guardian*, interviewing pale, interesting young women about their singing careers. Yet here he was, a junior reporter for *Caledonia News*, in the middle of nowhere watching the rain.

Three hours earlier, in the warmth of the Inverness office, his boss had approached him with a beaming smile waving a press release. Simon's heart had fallen the moment he saw that grin. Last week that grin had sent him to report the mysterious disappearance of several dustbins in Nairn. The week before that he'd used three years of university education to report on late-blooming daffodils. The resulting headline had been predictable. 'Fury as daffodils fail to bloom.' His boss was particularly fond of the word 'fury'. Simon wondered if he'd bought a job lot of the word cheap somewhere, as he used it at every opportunity. 'Fury as bins go missing.' 'Fury as nothing much happens.'

This particular press release had brought him here, parked outside some crumbling edifice of a stately home that looked as if it had just escaped from a Hammer horror film. On the lawn opposite the house a cherub stood above a pool smiling and peeing endlessly. His car had misted up and there was a faint smell of slowly decaying fairy cakes, the origin of which was a four-month-old sponge cake that had rolled under his seat but which he had never been able to find.

At last a rotund man with a big moustache came running from the door of the house of horror, carrying a big white sign with alarmingly red letters on it. He was being pursued by a younger, powerful-looking blond woman who was surrounded by a swarm of tiny braying dogs.

Before he opened the car door he heard the woman shout, 'Please don't make a fool of yourself again, Charles.'

Simon was stepping out of his car when he saw the small army of dachshunds charging towards him with the sole aim – he could see it in their eyes – of tearing him limb from limb. He hurriedly closed the door. For the next few minutes the couple fought to control the animals. Every few seconds a dog would leap at the window, all teeth and wild eyes, only to fall back to the driveway. After a few moments the woman managed to corral the beasts and drag them, yapping and snarling, back into the castle.

Now it was safe, Simon stepped out of the car and instantly felt the chill of rain sinking into his clothes. He saw a large hand thrust towards him.

'Charles Purdey, young man. My life is being made a living hell by the RSPB spying on me. You probably think of the RSPB as a group of harmless cranks, feeding twittering birds and that sort of thing. Well, let me tell you, I don't get a moment's peace. They are up and down the roads of my estate almost constantly and they have *absolutely no evidence* I've done anything. And furthermore—'

The reporter pushed a microphone in the general direction of the enraged aristocrat, whose diatribe against the injustices heaped on him by the stormtroopers of the ornithological world seemed unending.

When Purdey finally drew breath, the reporter got in a question.

'So it's a campaign, is it?'

Purdey thought for a moment and suddenly his eyes twinkled. 'Why yes, I suppose it is. Great idea. I'll start the campaign against the persecution of landowners. It's the CA—'

The reporter looked up from his notes and read back to Purdey. 'I think the CAPL.'

Purdey beamed at the young man. 'That's it – the CAPL. Got a ring to it. Prince Charles can be the patron.'

Blimey, Simon thought. *Maybe this is an actual story, at last!*

'So Prince Charles is the patron of the organisation, is he?'

Purdey hesitated. 'Well not yet but I'm sure he will be. He's patron of all kinds of things and we went to Gordonstoun together, you know.'

Simon tried to summarise what Lord Purdey had said. 'So the RSPB are persecuting landowners?'

Purdey grinned. 'That's it in a nutshell. It's all these lefty vegans, you know. They just don't understand the countryside.'

Lefty vegans, Simon scrawled in his notebook. *What a quote!*

'Can we have a photograph?'

Purdey grinned at the young man. Moments later the laird was grinning like a baboon beside his large white sign, now hammered into the ground. It read:

RSPB NOT WELCOME HERE

SUMMER

CHAPTER 5

The ragged west coast of Scotland has been shaped by countless years of battle with the Atlantic Ocean. What remains after this ceaseless struggle is hundreds of miles of coastline characterised by long fjord-like sea lochs, where the sea has invaded the land – and myriad islands marooned by the conflict. The Isle of Morvan is a small irregular island sitting across a few hundred yards of sea loch that separate it from the mainland. Geographically, the Minch of Morvan is an insignificant stretch of water crossed in less than five minutes by the little car ferry that plies back and forth carrying residents and tourists alike on to the island. In the summer, the *Maid of Morvan* – as the ferry has been christened by the Highland Council who run the service – shuttles continuously between the slipway on the mainland and the slipway on the island. Crossing to Morvan costs you £8.40, making this the most expensive ferry trip in Great Britain when measured by distance travelled per penny. This is a source of both great pride and irritation to the 400-odd folk who live on the island.

While the ferry crossing may be short, it has a huge effect upon the handful of square miles of wild landscape that constitute the Isle of Morvan. It makes these few square miles into an island, and that, on the west coast of Scotland, makes a big difference. The islands of the west coast all possess different personalities. In the south there is mild-mannered Arran, offering holiday homes to the citizens of Glasgow. Moving north there is the Isle of Skye where the romance of the jagged Cuillin mountains draws in thousands of tourists every year. Further north still there are the vast expanses of open moorland and endless windswept beaches of the Outer Hebrides. Off the northern tip of Scotland lie the Orkneys with their ancient cathedral and the bones of ships deep in Scapa Flow. Keep travelling north and you will eventually reach the Shetland Islands, which owe more to Viking ancestry than Anglo-Saxon or Gaelic, and boast, in their short winter days, the great Viking fire festival of Up Helly Aa.

Together these islands cluster off the coast of Britain like people sitting beside a table in the evening light. Morvan, too, has its place at the table. It is far from the largest island, and it is not the longest or the highest, but the folk who live there will tell you it is like nowhere else.

The moment the small ferry pulled away from the mainland Angus felt himself relax. Sitting in the Land Rover with Rory beside him, as they headed across the Minch to Morvan, it was as if he had left the world and its worries behind. They had been late that evening getting away from Inverness. Angus sometimes found his job recording the proceedings of council meetings onerous. That day it had been endless, the council split down the middle, neither side willing to yield. The problem had been the route of the new city bypass. There was the direct route, through an old wood that had protected status, and the longer, more expensive route, through farmland. One side had been swayed by economics, the other by the desire to protect the natural environment. Angus had sat there for four hours, recording the argument to and fro, as his shirt slowly stuck to his back with sweat and his eyes grew tired from taking notes. Hours ticked by. It had ended – as he knew it would – with Gerald McCormack rising to his feet and extending a skeletal finger towards those wishing to protect the wood. Others had heeded his warning of dire financial consequences should the longer route be considered. In the end the council always put cash before nature. The wood's fate was sealed.

It was well after dark when Angus shouldered his rucksack, turned on his head torch and began following Rory up the track towards Glen bothy. Though the track was steep and his pack heavy, it was good to feel the rough ground beneath his boots and catch the scent of the pine trees. Robbed of sight by the darkness, his other senses seemed more acute. At the top of the hill, just before they left the forest and stepped out on to the Black Moor, an owl hooted close by. He turned and saw the lights of its eyes reflecting his torch beam back at him, two glittering discs in the dark.

Rory heard it too. 'Tawny owl, that.'

Twenty minutes later they reached the river that marked the boundary between the Muir and Purdey estates. Angus swept his torch beam out across the Black Moor. This side of the bridge was treeless and here a faint moon gave a veiled impression of the landscape with the outline of the hills showing black against the starlit sky. On the far side, the oak forest would swallow them and the only light beneath the canopy would be their torch beams. The river foamed high as they approached the old wooden bridge, which they could hear creaking and groaning as the floodwater rattled its timbers. Angus looked down into the black, peat-filled water, frothing against the bridge supports.

Rory leaned against the rail of the bridge, shining his torch down into the water. 'I wouldn't fancy going in there tonight.'

Angus shivered at the thought. 'Aye, min. I'm with you there.'

The bridge gave a sudden groan and Angus felt the boards move beneath his feet. They both hurried to the other side.

A thin drizzle had started and they paused to take waterproofs out of their rucksacks. The island was often drenched in a fine mist of rain that could fall ceaselessly week after week, sometimes so fine that folk would hardly notice until, suddenly, they'd realise they were wet through. It was the kind of sneaky rain that made Angus wonder if it was really worth the effort of putting on rain gear.

Rory watched as Angus zipped up the cagoule he'd bought only a few weeks ago. 'How's that new jacket of yours?'

Angus raised his arm and admired the sheen of the new fabric, noticing with satisfaction how the rainwater was sitting in beads on the surface of the material. 'Aye, seems pretty good, though I've not been out in a real downpour yet, ye ken.'

Rory fastened his old jacket around him. His factory job meant new gear was a luxury he couldn't afford. 'What did you pay for that again?'

Angus was embarrassed to admit how much he'd paid, having an Aberdonian's reverence for frugality. 'Ah well, nae far short of 400 pounds.'

Rory whistled. 'Bloody hell, for that kind of money you better not get it wet.'

Angus zipped himself into his jacket and secretly enjoyed the smell of the new nylon. This was probably the best jacket he'd ever owned, but he had decided that at his age he could allow himself the odd luxury now and then, even if he felt a twinge of guilt at the cost. They had only walked a few hundred yards when the skies of Morvan decided to open and the fine rain turned to drops so big that they thundered down like bullets bouncing off the forest floor.

Rory hunched forward, trying to shield himself from the downpour. 'Oh bugger.'

Angus stayed silent, not wanting Rory to think he was gloating over the protection his new jacket offered. They hurried on and soon found themselves standing before the small stone shelter of Glen bothy. A pale yellow light showed in the windows and their torch beams picked out smoke drifting from the chimney.

'Someone's in,' Rory announced, spitting out rainwater that had run into his mouth.

Angus had been looking forward to a quiet night. 'Aye, like as not it'll be full of louns an' quines from some uni mountaineering club getting drunk and smoking pot.'

Rory laughed. 'Angus Sutherland, you're turning into a miserable old git, do you know that?'

Then Angus laughed too, and they stood for a moment watching the rain running in rivulets down the grey slate roof into the old guttering where it spouted from cracks and cascaded on to the stone walls. Angus wondered how many rainstorms the old bothy had withstood over the years.

Rory pushed open the heavy wooden door and together they stepped into the welcome warmth of the bothy. The steam of cooking pots filled the air, and the light from the flames of the wood fire danced on the wooden wall panels of the ancient cottage. A man and a woman were sitting in front of the fire. Clearly they weren't students, but to Angus they appeared to be people who had not long left university. The young woman was tall and slim with dark hair and had an unmistakeable air of authority about her.

She smiled and put out her hand.

'Hello there. You'll be Angus and Rory. You better come in and join the hooley.' She sounded well educated, but her voice carried the kind of soft Northern Irish accent that always made Angus think of Guinness and warm cheese. 'I'm Sinead,' she said, shaking Angus's hand.

Angus was puzzled as to how Sinead had known who they were, but then he spotted the thin figure of Brian sitting quietly in the corner, as ever, and the mystery was solved.

Sinead was obviously used to organising everyone. Within minutes, she had everyone introduced, the sleeping arrangements sorted, and Angus and Rory slowly drying out before the fire.

Sinead turned to Rory. 'Is it you two who have the old Land Rover?'

Rory laughed. 'Yes, that's us. You don't miss much, do you?'

'That's my job. I'm Tony Muir's estate manager,' she said, taking a sip of her wine.

Angus was surprised. Most estate managers he'd met were crusty old Highlanders, not young Northern Irish women.

Rory put it into words. 'You're an estate manager? You don't look like an ordinary estate manager.'

'You mean I'm a woman?' Sinead smiled and her dark eyes

sparkled in the candlelight. 'Ah feck, this is no ordinary estate either.'

Rory nodded slowly. 'Yes, it's why we like coming here. It's not like Purdey's estate, with treeless hills and a bloody grouse moor. It's nice to see a bit of natural forest for a change.'

Sinead brightened at Rory's words. 'Aye, we like the oul place too.' She looked over to where one of the men was tuning his guitar. 'Is that not the truth, Hector?'

Hector, a man in his forties with a closely shaven head, smiled and waved, and even Brian nodded his approval.

Sinead guided them to a bench beside the fire. 'Youse two are here a lot, aren't you now?'

Angus was surprised to think that the estate manager knew of their existence, and was reluctant to admit how often he used the bothy. 'Aye well, we're here now and again, ye ken.'

Sinead looked at them with mock surprise. 'By God, there must be someone with a bloody Land Rover just like yours, then, so there must, cos I see it here most weekends, so I do.'

Rory had no such reserve. 'Yes, we're here quite a bit. There's lots to do here and plenty of wildlife to see. At least there is on this side of the river.'

'So where's youse from, Rory?'

Rory, evidently deciding that the evening had begun, opened a can of beer. 'We both live in Inverness, but I'm from Liverpool. Angus is from Aberdeen.'

'Liverpool,' Sinead chuckled, 'the fecking capital of Ireland. So you're Irish like me. What's Inverness like? Every time I go there it gets bigger.'

'Aye,' Angus said, 'it's fair growing at some rate.'

Rory became animated. 'I couldn't believe it when I moved to Inverness ten years ago. I grew up by the River Mersey. It was full of chemicals and so much oil the water would catch fire if you chucked a match in.' He took a sip of beer and wiped his mouth with the back of his hand, warming to the subject. 'Then I moved to Inverness and the town had a salmon river running through it and there were otters. Once I saw an osprey take a fish from the River Ness. And I could walk to the pub following seals swimming upstream. I couldn't believe it.'

Angus shook his head sadly. 'Nae so many salmon these days. Fished out.'

Brian shuffled silently across the room and headed for the single other room the bothy possessed. Angus called after him. 'You off to bed?'

Brian didn't turn to answer but called over his shoulder. 'Aye, that's me away.'

Rory looked at his watch. 'Half eight.'

Sinead glanced at Angus quizzically. 'You'd think he'd stay up for the craic.'

Angus whispered to her. 'That's Brian. He disnae do craic. Always in bed early, but he'll be away before dawn.'

The Northern Irish woman looked mystified and stared down into her whiskey. 'Takes all sorts, I suppose.' She called over to Hector, who was still tuning his guitar. 'Hey, Hector, are youse gonna play that thing?'

Hector smiled and scratched his shaven dome. 'Patience now.' He went back to his tuning with renewed determination.

Angus poked the fire and it responded by sending sparks up the chimney. 'I don't think I've ever seen an estate worker in a bothy before. Odd that, really.'

Sinead's dark eyes sparkled. 'Youse'll not be seeing any Purdey staff up here, to be sure. These hills are where they go to work. When they've finished they go home to their centrally heated houses. They'll never understand why anyone would want to spend time up here in a draughty old bothy when they could be at home.'

'How come you're here, then?' Rory asked.

'We're up here to bash bracken,' she explained.

Angus looked at her puzzled. 'I do nae ken ... '

Sinead stood up and walked to the table to get her whiskey flask. 'It's the pearl-bordered fritillary we're here for.'

Angus was even more puzzled.

'It's a butterfly,' Rory said.

Sinead nodded. 'Well done, I'm impressed.'

Rory looked pleased with himself.

Sinead sat back down and offered her flask around. 'It's good stuff. Tullamore Dew, proper Irish whiskey.'

Angus took the flask and poured a measure of the golden liquid into his plastic mug. 'So what's bashing bracken got tae do wi' butterflies?'

Sinead took a sip of her whiskey and savoured it for a moment. 'We've fenced off that part of the forest to keep the deer out but that means the bracken grows too high.'

Rory nodded. 'I see, and that stops the dog violets growing because the forest floor is too dark.'

Angus couldn't see the connection between butterflies and violets and shrugged.

There was a sudden burst of sound from Hector's corner, and

he started singing. 'We're bashing bracken for butterflies … '
He managed the line twice before he gave up trying to find
a line to follow it.

Sinead wrinkled up her nose at the tuneless sound. 'Jasus,
Hector, youse sound like a cat screaming.'

Angus felt left out of the conversation and was increasingly
ashamed of how ignorant he was. As club president he should
have been more knowledgeable about the wildlife of the mountains. He realised that, for all the years he'd wandered the hills,
he knew very little about the creatures that inhabited them.
Rory had only a fraction of his experience but had a far greater
understanding of the environment. Angus felt as though he had
been blind for all these years.

With more enthusiasm than talent, Hector began to sing a
song about the wild moors and the hills. No one seemed to mind.

One of the candles died and Sinead got up to replace it.
'What is it you do yourself, Angus?'

'I am chief recording officer for the Highland Council,'
Angus said, expecting Sinead to be impressed.

Sinead scowled and spoke with venom. 'Those feckers! I've
nae time for them. That council always votes for the landowners.
Anything for jobs an' a few quid.'

Angus looked at her in shock, surprised at the estate
manager's vehemence. 'The council has to support the local
economy. People have to work.'

Sinead stabbed her finger at Angus. 'Aye, people have to
work. But every time there's a feckin' application for rewilding
or a rich landowner wants to do something, they just cave in.'

'She's got a point, Angus,' Rory said in a gentler tone.

Angus wasn't going to let the reputation of the council be degraded so easily. 'That's not always the case, yer ken. And, in reality, we can't just turn back the clock and let the place go wild. It's just not—'

Before he could finish his sentence, Sinead thumped her mug down on the rough wooden bench so hard that Hector stopped torturing his guitar and glanced over at her. 'For feck's sake, they're blind to what you've got here. The Highlands could be an incredible environment. A showcase for nature. Instead of that you let people bulldoze the place to the ground, an' put feckin' wind farms all over the place.'

Rory nodded furiously. 'Not to mention what we do to the sea. Salmon farming is an environmental disaster. And there's scallop-dredging destroying the seabed.'

Angus could feel himself being backed into a corner. 'Yer cannae turn the whole of the Highlands into an ecological museum.' Even as he said the words, he wondered if he truly believed them – wondered if he'd internalised what he'd heard so many times before in council meetings.

'Not *everywhere*,' Sinead said. 'But there could be some part of the Highlands where we let the forests come back and give nature a bloody chance.' She deflated like a balloon. 'You'll have to forgive me. I get so fed up of nothing changing. We never seem to get anywhere.'

As the evening progressed Hector's singing became less tuneful and Sinead, sipping more of her Dew, grew more talkative. 'I wouldn't work on a sporting estate. Tweed and blood sports aren't for me.'

'But this estate does shoot deer. You do have stalking,

don't you?'

Sinead nodded. 'I shoot deer myself. We have to – there are too many. We have to control them and there is no other way.' She paused, and then, whispering as though there were spies outside the bothy: 'Rewilding, that's what I'm working on. It's our long-term plan.'

Angus was curious. 'Do yer really think it could be done?'

'The reintroduction of native species?' Rory said. 'Beaver, lynx and the like, for sure. Imagine what it would be like here if there were lynx in these woods.'

Angus shook his head. 'Dangerous, dangerous road tae go down, that. They're big animals, lynx.' He looked over his shoulder as though one might be creeping up on him at that very moment.

Sinead was keen to reassure him. 'There have always been lynx in parts of Europe. Not one recorded attack on humans.'

'Sheep, then.' Angus poked the air with the stem of his pipe to emphasise the point. 'You cannae tell me that they widnae attack sheep.'

Sinead put up her hands in mock surrender. 'OK, youse have me there, but only around five per year. More than that get killed on the roads.'

Rory was warming to the subject now. 'I've been reading about reintroducing the lynx. But what about the native wildcat? Won't they compete for food?'

Sinead shook her head, clearly pleased to be able to talk about her passion. 'No, not at all. The lynx mostly prey on roe deer, although they might take a red deer. Wildcats take smaller prey – rabbits, voles and the like. They live alongside each other in Europe, so they do.'

'So, what about here on Morvan? Do you think this estate would sustain lynx?' Rory looked excited.

'Aye, maybe they could live here,' Sinead said thoughtfully. 'As long as they could wander over a wide area.'

Angus put down his pipe. 'Now there's the problem, yer ken. Can you imagine the likes of Purdey agreeing to lynx on his land?'

'It shouldn't be down to men like him,' Rory snarled.

Sinead nodded. 'Landowners have too much power, do they not?'

Slowly the conversation turned to revolution. Angus tried to argue for the moderate course but eventually realised that the hour was late, whisky had been drunk, and this was a time for him to simply sit, smoke his pipe, and let the next generation talk.

If you had been outside the bothy, leaning quietly against the old stone wall as the rain ceased, you would have seen the moon peering down from behind the parting clouds and filling the glen with pale, silvery light. If you had made no sound, and had the patience to wait a while, you would have seen the brindled wildcat steal silently from the forest, its eyes fierce gold in the moonlight. You could have watched as it took a vole, black and wriggling, from the edge of the path and then vanished into the secret forest, like a lost memory. At length, you would have seen the light in the bothy windows fade as those inside blew out the candles and turned in to sleep, and you would have felt the falling silence consume the glen.

***

Rory awoke to sunlight filtering through the bothy window and a metallic rattling close by. He crawled out of his sleeping bag and found Angus stirring his porridge in the main room of the bothy.

Rory stretched and tried to orient himself. Hector was still snoozing and Sinead's sleeping bag lay empty on the floor.

Rory looked round the bothy. The cheerful fire of last night had died down to a mound of grey ash. 'Brian away?'

Angus handed Rory a cup of tea. 'Aye, long gone.'

In daylight the bothy was once more just an old draughty cottage festooned with cobwebs and dust. The door rattled and then burst open and Sinead almost fell through the opening.

'Och, fir feck's sake, my head.' She put her hand to her forehead and stood for a second swaying gently.

Angus made the Northern Irish woman a cup of tea. She consumed it – and two more – before the colour returned to her cheeks.

Rory spooned some porridge into his plastic bowl and sat down on the bench next to Sinead, who was rapidly recovering from her hangover. 'What part of Ireland do you come from?'

'Just a wee village outside Belfast, Dundrod. Blink and you'll miss it.'

Rory finished his breakfast and began packing his rucksack for the day ahead. 'Beautiful place, Ireland. My dad took me there fishing when I was fourteen. It was amazing – the air was full of butterflies and insects and the fields full of wild flowers.'

'Doesn't sound like Belfast to me,' Sinead said.

'No, it was southern Ireland. A place called Mullingar. It was like going back in time.'

Angus was gathering the pots to wash them. 'Whit do you mean?'

Rory was carried way with his memory of that holiday long ago. 'In Merseyside, where I grew up, you didn't see insects like that, or flowers. It was all intensive farming and pesticides. My dad said it used to be the same as Ireland back home when he was a boy. I wish we could see it like that again.'

Sinead stood ready to head out for the day, fastening her dark hair behind her head. 'That's what we're doing. Trying to bring the butterflies back. Come and help us, why don't you?'

Rory turned to Angus. 'Yes, why don't we give them a hand?'

Angus had been looking forward to his walk all through the working week and wasn't keen to give it up. 'Ach, I'm nae sure aboot that, yer ken.'

'Come on, we can walk tomorrow,' Rory insisted.

Angus was reluctant to give up his freedom, but it occurred to him it might be good to make a small contribution and help the pearl-bordered whatever-they-were, for one day at least.

Aye, all right, why not.'

***

Sinead led Angus, Rory and Hector down beside the river and into a gorge that Angus had never noticed before. Around them the wooded hillside rose steeply while at the foot of the gorge the river that divided the two estates gurgled and foamed, swelled by last night's rain. They climbed a fence that Sinead told them was to keep out the deer and headed into the wood. The bracken and small trees and bushes made it difficult to walk at any pace here, and Angus realised that this was the forest regenerating after the exclusion of the deer.

Sinead stopped and bent down. 'See that white flower? That's the dog violet I was telling youse about yesterday. That's where the butterflies lay their eggs.'

Hector made some notes. As Sinead walked through the wood she would stop now and again and point out a flower or plant – things Angus realised had been there all his life, but he had walked blindly past. It was as if there was a whole different world he'd never seen before.

Sinead pointed to a small orange butterfly floating past. 'That's one there. Pearl-bordered fritillary.'

Rory peered at the little butterfly. 'I don't think I've ever seen one before.'

Hector scratched his bald head. 'They are pretty rare. Perhaps they'll make a comeback.'

They spent the day breaking down bracken to get more light to the woodland floor. Hector explained that the aim was to clear about a third of the bracken as, ideally, the habitat would have about one third shade and two thirds light.

Slashing down the bracken with their walking sticks was hot work, and at first Angus thought it a pointless exercise, but after a couple of hours he noticed that they had made an impact on the vegetation.

Sinead called a halt to their work. 'I think that's enough for today. Come on wi' me, let me show youse something.' She led them down to the riverside, pushing her way through bracken and bushes as they went.

At last they stopped and Sinead pointed across the river. 'There, what do youse think of that?'

Angus followed her gaze and there, on the opposite side of the river, was a small copse of twisted trees, their reddish, scaly barked limbs distorted by the wind. There must have been around a dozen trees, some with their branches so misshapen they reached out and dipped into the river.

Rory looked over at the trees, appreciative. 'Scots pine! I had no idea there were any here. Not this old.'

'Neither did we, not this far west,' Sinead said quietly. 'I think maybe they're a relic of the great forest of Caledon. They're old, that's for sure. Maybe just a few fragments made it this far.'

Angus had seen Scots pine in the Cairngorms, and knew there was a remnant of the ancient forest near Torridon. He'd heard it was being conserved somehow but knew nothing about it, really. 'Ken we've been walking past here for years. Never a' thought that those trees were here.'

Sinead laughed. 'Ye never know what youse'll find until you look.'

Angus realised that he had missed so much in the landscape he travelled by not looking. 'Aye, I ken that fine.'

***

That night the bothy was quiet. Sinead and Hector had gone, taking the sounds of the guitar and – perhaps mercifully – Hector's singing with them. As the evening drew on the sky clouded over and soon sent down a heavy, cold rain, making Rory grateful for the fire that kept them warm and the walls that stood firm against the deluge. Every now and again they would hear the wind crashing through the trees to pound against the roof like an angry giant. The wilder and wetter the elements outside, the warmer and more secure they felt inside. The rain continued without pause.

The next day, Rory, Angus and Brian stood, battered by rain, staring at the few pieces of split timber that remained of the bridge. The river boiled dark and forbidding like some angry creature railing against the confines of the riverbank. The daylight had gone by now and the broken timber and swirling water looked even more threatening in Rory's head torch beam.

'It's the only way oot,' Angus said. 'We could try walking out over the hill but if we did that we'd have tae ford the Black Water and I widnae give us much chance of doing that in this weather.'

Rory peered into the churning pool. 'How deep do you think it is?'

'Nae more than three feet I'd say. We can wade it.'

'Oh bugger.' Rory had waded rivers before and hadn't enjoyed it.

Angus put on his confident mountaineering club president voice. 'Probably only get our feet wet. We'll do what we did in the Scouts, yer ken – we'll all link arms in a circle, and that way we support each other as we cross. It'll be aw richt.'

Rory didn't see much all right about it. 'Are you crazy? We'll get washed away.'

Brian also looked doubtful, but Angus chuckled. 'Away, man. I've done this before. Come on, link arms.'

They linked arms in a small circle. A newspaper headline flashed into Rory's mind: 'Hillwalkers Washed Away in Tragic Drowning Horror'. He did his best to push the thought from his mind but it was still there as he stepped into the river and felt the chill of the water entering his boots.

Angus shouted to make himself heard. 'Now just hold on tae each other and we'll be fine.'

The trio edged deeper. Now the water was over Rory's knees. 'Maybe we should go back.'

'Just keep moving, we'll get there,' Angus yelled as the water reached his waist.

Rory glanced at the far bank. It seemed awfully far away. He stepped into deeper water and sank to his chest. 'Bloody hell.' Then he realised his feet were no longer in contact with the riverbed, and over the writhing water he yelled, 'Angus! Are your feet on the bottom?'

Angus spat out some water that had splashed into his mouth. 'No, are yours?'

'Oh bugger,' Rory spluttered.

Both men turned to look at Brian. Brian didn't speak. He didn't need to; the look of unbearable sadness that passed across his face was the only response they needed.

Moments later, they were moving. Trees passed by overhead with remarkable speed, they passed a semi-submerged rock, and for a few seconds spun in a small whirlpool as if performing an aquatic version of the Highland fling. The river grew rougher still and Rory found himself under the water for seconds at a time. As he broke the surface, he took huge gulps of air, the beam of his head torch catching the trees overhead. Rory realised they were in the gorge where they had been bashing down bracken the previous day. Then he heard it: a deep-throated rumble, coming closer with every second.

'What's that?' Rory spluttered, breathless.

Brian didn't often speak; it was as though he had been issued with a limited number of words and didn't want to waste them. At this particular point in time he chose to use some of his precious quota. 'That'll be the waterfall.'

'Hold tight!' Angus yelled.

Rory thought the suggestion was unnecessary, as most people about to be catapulted into space over a ten-foot waterfall will cling with unbridled enthusiasm to anything they can lay their hands on.

Then Rory remembered the old Scots pine with its branches almost touching the water. Moments later his head torch beam picked out a twisted branch reaching out to them like a giant trying to pluck them from the water. Now that the waterfall was close its roar filled Rory's ears. He clung to both of his companions and knew their only hope was to catch the passing branch.

He tried to free his arm, but Angus just gripped tighter. 'Angus, let go.' A look of horror passed across Angus's face, and Rory yelled again: 'The branches!'

Angus and Brian turned and saw the tree. Angus let go of Rory's left arm and vanished into the darkness with only Brian for support.

Rory lunged for the tree. He touched it with his fingers but the bark slipped through his wet hands and was lost in the darkness.

Shit, shit, shit, he thought, panic rising. Then a second branch loomed out of the night. He lunged again and this time he caught it. He pulled Brian into the tree with his right arm and Angus followed. They edged along the branch and suddenly Rory felt the glorious sensation of solid ground beneath his feet. Moments later all three of them lay gasping on the riverbank.

Rory heaved himself from the ground and stood there, shivering, with water cascading from his clothes. 'Must have been a tough life in the Scouts in your day if they crossed bloody rivers like that.'

They all laughed, mostly from relief at escaping from the clutches of the river.

Angus looked pale in Rory's torchlight, but he smiled. 'Aye well, at least you've got your river-crossing badge.'

***

Charles Purdey was dozing beside the fire in the great hall of Castle Purdey, a copy of the *Financial Times* spread across his ample midriff, which rose and fell as he snored. Tabatha sat opposite him flicking through Twitter on her phone. She heard footsteps running down the hall outside and turned in surprise just as Donald and Hamish burst through the ornate wooden doors.

The dachshunds at Tabatha's feet erupted into a cacophony of barking and Purdey woke with a start. 'What the hell?'

Hamish was red-faced with excitement. 'Poachers, sir!'

Donald nodded. 'Aye, he's nae joking. Lights by the river.'

Purdey rose in fury, like his ancestors had done when they heard the redcoats were on the moor. 'By God, I'll not have this.'

Tabatha got to her feet as well but Purdey turned to her. 'You'd better stay here, m'dear – this could be dangerous.'

Tabatha sighed. Nothing this exciting had happened on the estate for years and she certainly wasn't going to miss it. 'Don't be ridiculous, Charles.'

Purdey capitulated at once. It was useless to argue with

his wife.

The Land Rover screeched to a halt and Purdey, Tabatha, and the two ghillies tumbled out on to the gravel track. Below them, down by the river, lights bobbed about in the darkness.

Purdey locked his shotgun breech closed and prepared for battle. 'Hamish, get the searchlight on them.'

Hamish reached up on top of the Land Rover and swivelled the big mounted searchlight into position. Tabatha stood beside her husband, a stout walking stick in hand.

Hamish switched on the light and the scene beside the river was instantly illuminated.

Purdey gasped in astonishment. 'Good God. There appear to be three naked men by the river.'

'So there are, Charles,' Tabatha mused. 'We don't normally have those at the bottom of our garden.' She took a cigarette from a silver case, lit it, and inhaled deeply. 'I suppose it's better than fairies.'

CHAPTER 6

As the old Land Rover rattled through the deepening night, following the tarmacked shore of Loch Ness, the moon tiptoed across the water painting its unfathomable surface with light. Rory was dozing in the passenger seat and Inverness had slipped into its Sunday evening nap by the time they reached the outskirts of town. Angus felt himself sucked into suburbia by the neon street lights as they pushed back against the glowering night. The Land Rover coughed and lurched as he pushed it around the new roundabout and on past the sports centre, with its illuminated circle of Lycra-clad runners, and beyond past the 1960s practical tedious rectangle of the Highland Council offices. The dull ache in his legs reminded him of the walk he had just taken through the hills, and for a few seconds he again saw the path beside the loch, smelt the acrid woodsmoke of the bothy and felt the wind against his face. Angus was crossing the threshold between two worlds, returning to his ordinary life in his comfortable semi in an ordinary street. That thought drained the air from his lungs and the car sagged a little as it turned in to the estate.

Rory was still sleeping when Angus brought the Land Rover to a halt outside the block of flats beside the river that was home to Rory and Jen.

'Sleeping Beauty, we're here,' Angus said gently.

Rory spluttered awake and spent a moment staring at the street lights. 'I must have nodded off.'

Angus laughed. 'You've been snoring since we got off Morvan ferry, yer ken.'

Rory eased himself out into the street and walked to the back of the vehicle to pick up his rucksack. He shouldered the pack, which dripped water down his back and on to the pavement.

Rory reached into his pocket and produced a crumpled note. 'For the diesel.'

Angus looked at the sodden banknotes dubiously. 'Maybe settle up at the club.'

Rory nodded and made to walk towards the flats. He turned to Angus. 'I don't think they recognised us. How could they?'

Angus sighed. How would the council react if it became known that one of its officials had been spotted nude on a riverbank with two other men in the middle of the night? He put the Land Rover into gear and left Rory watching him drive away, dripping steadily.

***

Angus paused stiff-legged outside the kitchen door, his wet clothes dangling in a plastic bag. The light from the window spilled out on to the patio. He took hold of the door handle, took a deep breath and stepped into the kitchen. The sudden shock of the electric lights left him blinking. The air was stifled by the smell of chicken soup steaming slowly in a pan on the stove. There was a sheet of newspaper spread out on the tiled floor awaiting his entrance; he stepped on to it and waited. A moment later, Laura emerged from the lounge as she always did when he returned from the hills. She was a tiny woman, the thickness of a matchstick and just under five feet tall, but her diminutive size didn't stop her filling the room.

'Had a good time?' It wasn't a question – it was an accusation.

Angus held up the plastic bag full of wet, heavy clothes. 'We had to wade the river, yer ken.'

He decided to edit out the highlights of the day. Best not to mention almost drowning and, most importantly of all, not say anything about the indignity of being found naked on a riverbank.

Laura took the bag and stared at it in disgust, as if the black plastic contained a severed head. 'Did you, now?'

Laura had that look on her face, the one she used when she knew Angus wasn't telling her the truth. He tried to look casual. 'Aye, the bridge was washed awa."

Laura sighed heavily, like a high court judge listening to the accused lying for the thousandth time and hoping the jury would get the message. 'I made some chicken soup.'

Angus tried to look pleased. 'That's just what I need.'

His stomach churned at the thought. Laura had an amazing ability to cook the most revolting chicken soup ever made. How she got such a peculiar, insipid taste into the simplest dishes Angus could never work out. Laura did not cook for pleasure – her cooking was a duty, a chore that showed she was a good wife. The food she prepared was not to be enjoyed, it was to be consumed with a kind of reverence for her abilities as a house-wife. Angus regularly had to endure an array of disgusting food that Laura took great pride in cooking. They dined on tripe, and home-made oxtail soup that took hours to prepare and tasted like a gelatinous gloop. Laura also cooked a dish of some kind of meat that had to be weighted down and boiled for hours. Her cooking was an endurance test.

She would proudly reveal her latest creation of something last eaten in the Dark Ages and announce, 'The butcher said no one eats this any more.' Angus would reflect that no one uses leeches to cure fever any more but we wouldn't want to reintroduce that practice.

The kitchen walls sweated with the steam from the soup and the windows wept condensation. Angus looked around the kitchen with a growing sense of unease. Nothing good ever happened here and the white-tiled walls reminded him of childhood trips to the dentist. This was Laura's headquarters, where she planned what would happen during the week and decided if Angus would be allowed out to play at the weekend.

Angus peered at the pan of soup, watching the lid rattle as though something inside were about to escape. 'Wonderful,' he said, although he knew it wouldn't be.

Laura pointed a stubby finger at the newspaper neatly unfolded on the tiled floor. 'You stink of smoke and sweat.'

Angus stripped off the rest of his clothes until he was left standing in his underwear, feeling like a skinned rabbit. 'Did Christopher phone?' he called after her.

'No.' Her response was terse. Christopher, their son in his early twenties, rarely called.

'He might not have signal where he is, ken,' Angus said.

Christopher was climbing somewhere in France, probably Verdon, and Angus knew the steep-sided valleys often left big areas bereft of phone signals. He also knew that Christopher was probably reluctant to get another lecture from his mother.

There was a metallic click from the utility room and Angus heard the washing machine start up.

Laura returned to the kitchen, vigorously towelling her hands as if she'd just slaughtered a goat and was trying to rid herself of its blood. 'I met Jean Philips in Tesco.'

Angus knew what was coming next but tried to sound cheerful. 'How is she?'

'Their Peter's doing terribly well now. He's an accountant in London.' Laura paused to allow that information to sink in. 'Good salary, he and his wife ... What's her name?'

Angus had pulled on pyjamas and slippers by now. 'Annabelle.'

He had the feeling that Laura knew perfectly well who Peter was married to but was forcing him to recall their wedding as a way of bringing their own son's continued single status into sharper focus.

Laura nodded and a thin smile crossed her lips. 'Oh yes, Annabelle. Nice girl. She's expecting, Jean says.'

Angus froze midway through fastening his dressing gown. Laura had dropped the grandchild bomb. Angus had introduced Christopher to climbing, thereby ruining Laura's life. Christopher spent his life wandering the climbing grounds of the world. Frequently penniless, normally unemployed and always in imminent danger of death, he seemed likely never to provide Laura with the thing she most dreamed of, the golden-haired gurgling grandchild. This, of course, was Angus's fault, and Laura would hold it against him until time itself crumbled into dust.

Angus knew there was nothing he could say that would save him now from Laura's stealthy, deliberate but inevitable bestowing of guilt.

Laura turned her attention to the chicken soup and stirred it vigorously for a long time, filling the kitchen with steam. 'Such a nice girl, Angela.'

The words entered Angus's skin like a thin blade. He waited, like a man on the scaffold, for the executioner's hand.

Laura sighed, a long, simmering letting out of despair. 'She was so patient waiting for him and then he had to go off again.'

'Well, he's young and he's doing so well at it.' Angus felt as though he were standing up in court trying to defend a murderer who, after killing someone on live TV, had two million witnesses against him.

Laura ladled out a bowl of chicken soup with the deliberate precision of a poisoner. 'Yes, I'm sure you're right. As long as he's happy.' She turned and walked out of the kitchen, throwing another sentence after her as she went. 'Don't let your soup get cold.'

Angus looked down into the steaming bowl. It *looked* good. He could see bits of chicken, rice, barley, carrots and onions. Why wouldn't it be good? He dipped in his spoon and tasted the first mouthful. It was quite good at first, almost pleasant, but then it came. He was sure he could taste it. There it was, an unmistakeable aftertaste of bitterness, as surely as if it had been laced with cyanide.

He forced it down and put the bowl into the dishwasher and noticed two wine glasses sitting waiting to be washed. Perhaps Laura had had two separate drinks of wine. Yes, that was it;

what other explanation could there be?

He headed to the lounge. On the way a faded photograph on the wall caught his eye. It was of a young man and a small pretty dark-haired woman standing beside a camper van. He barely recognised the slim young man with long, dark hair with his arm around the waist of the girl smiling up at him. He remembered that day, leaning against the old VW Camper that broke down in just about every way possible on every trip. That hadn't seemed to matter back then. They had got where they were going somehow. That holiday they spent the long summer days climbing on the rough gabbro of the Cuillin and bathing in the clear water of the pools beneath the cliffs. Angus stood for a moment, looking at the black-and-white photo of two people he barely knew.

She was still there, somewhere, that pretty girl with the ready smile, but the dust of thirty years of marriage had covered her over. 'Where did we go? I cannae mind.'

'What?' Laura was settled in front of the TV.

'Oh, nothing.' He sat down in his armchair in the lounge. Then he remembered the two wine glasses. 'Anyone call over the weekend?'

Laura was popping a chocolate into her mouth. 'No, nobody.' She chewed for a moment and then stopped. 'Why do you ask?'

Angus picked up the newspaper and opened the sports page but didn't read it. 'Oh, no reason, yer ken. I just wondered.'

Why two glasses? Then he realised that she was eating chocolates. *She had a box of chocolates.* He put down the paper and glanced at Laura, who was lost in the TV. Beside her was a box of expensive chocolates. No one buys themselves a box of chocolates – least of all Laura, who had become almost obsessively frugal. You are given a luxury like that by someone, an admirer perhaps. He pushed the thought away, the thought of Laura and someone, pushed it into a box in his head and closed the lid, just as he always did. Just as he had done for the last few years.

He had been sitting for a few minutes looking at the TV – not watching the images, just letting the colours flicker before his eyes – when a familiar face came into focus.

A dark-haired woman was leaning against a farm gate being interviewed by a tall young man with longish hair.

'Sinead!' Angus shouted in surprise.

Laura glanced at him, alarmed. It was rare for Angus to take much interest in the TV. 'What, Angus?'

On the screen, Simon Partington said to the camera, 'I'm talking to Sinead Callaghan about the upcoming application to introduce wild lynx back into the Highlands.' He turned to the Northern Irish woman. 'Many people may not have heard of lynx before. You represent the Back to the Wild campaign. Can you tell us a little about the lynx and why you think they should be allowed back into the wild?'

Sinead smiled. 'Lynx are large cats, a little like a leopard but smaller, about the size of a Labrador. They are native to the UK but became extinct here over a thousand years ago. We in Back to the Wild see no reason why the lynx could not be selectively reintroduced. Our report makes the positive environmental impact perfectly clear.' Sinead held up a thick folder.

Simon smiled at the camera and turned back to Sinead. 'Surely an animal like that would pose a threat to animals, perhaps even people?'

'Wild lynx already live in parts of Europe and Scandinavia. They can live perfectly well alongside people and are secretive so avoid contact with us at all costs. It is likely that there would be limited predation on sheep, but we think it important that farmers are compensated. Lynx are forest animals and would help keep the roe deer population in check.'

Simon looked at the camera once more. 'The Highland Council is shortly to consider an application for a licence to release lynx back into the wild here in the Highlands. But these plans are controversial and there are many concerned voices against them.'

The screen cut to a figure in tweed beside a shiny Land Rover Discovery. He was stroking a dachshund.

Simon struggled to contain his hair in the wind. 'Lord Purdey, you in the Campaign Against the Persecution of Landowners have strong views about the reintroduction of an apex predator.'

Lord Purdey looked grim. 'Indeed we do. We represent the farmers, the ordinary people of the Highlands who fear for their livelihoods should such a ferocious creature be roaming the hills and forests. There would be a danger to livestock and people too.'

'So did you start this campaign in response to the attempts to release the lynx?'

The dachshund Purdey had been stroking suddenly began to wriggle and then jumped out of his hands. 'The lynx are only part of it. We who live in the countryside are positively under siege. You ask any shepherd up here. Sea eagles are *decimating* the lamb population. Poor little things. Frolicking one minute, disembowelled in mid-air the next.'

Simon glanced anxiously down as the sound of growling could be heard. 'Sea eagles are a protected species, are they not?'

Purdey raised his finger and pointed to heaven. 'Now that's the problem. And people like the RSPB are spying on innocent landowners like me. Accusing us of all sorts of things when they have absolutely no evidence.'

The growling grew louder and was now accompanied by a tearing sound. 'The Royal Society for the Protection of Birds—' Simon began, squinting down at his trousers and looking worried.

Purdey glanced down. 'Will you stop that,' he commanded, but by now the small dog had its teeth firmly fixed on the journalist's trousers and was doing battle with them for all it was worth. 'I and many other landowners are sick of these wildlife bodies constantly accusing us of harming birds of prey when we've done nothing wrong. It's harassment.'

Simon was struggling to stay upright as the little sausage dog fought to rip off his jeans at the ankles. 'But raptors like eagles and harriers are going missing, are they not?'

Purdey surreptitiously tried to push the dog away from the reporter with his foot, perhaps hoping that the viewers wouldn't notice the altercation. But what the dog lacked in size it made up for in tenacity and the growling continued. 'I wouldn't be surprised if it wasn't these wildlife groups themselves doing it to give us landowners a bad name. A false-flag operation.'

The reporter was gyrating wildly by this stage, one leg being towed around by the furious dachshund. 'That's a serious accusation. Have you any evidence?'

The TV screen cut to a close-up of Purdey, who looked alarmed. 'Evidence? They can't prove it's *not* them, can they?' A sadness came into his eyes. 'These are all very well-meaning people who suggest these things. Most of them don't know anything about the countryside.'

There was a cry off-camera and the sound of tearing trousers.

Purdey tried to keep going. 'It's the vegans, birdwatchers, walkers ... cranks, basically. We people who live in the country-side have its interests at heart. I've been conserving and then shooting wildlife ever since I could walk. There's no way I'd stand for lynx being released, I can tell you.'

The reporter appeared on the screen again, red in the face and sweating. He tried to say goodbye over the growling and the ripping sounds and the scene quickly cut to a feature about the closing of public conveniences on the west coast.

Laura was staring at Angus, shocked. 'Do you know that Sinead woman, then?'

Angus nodded. 'Met her in Glen bothy yesterday. She never said anything about the TV.' He left out the detail about helping her bash down bracken. The less Laura knew about what he did on the hills the better.

Laura took a coffee cream from the chocolate box and chewed it slowly. Angus could tell that she was thinking, and nothing good ever came of that.

'Ridiculous,' she said at last, 'releasing lynx into the wild. Who'd want those things roaming around? Dangerous.'

'They're no dangerous,' Angus said. The words had flown out before he could stop them. He never argued with Laura; he'd realised how pointless that was years ago.

Laura stopped chewing and gave Angus a withering look. 'They're big cats, like leopards, even that ... what's her name?'

'Sinead.'

'Even *Sinead* said that.' Laura closed the lid on her chocolate box. 'Never mind. I'm sure Gerald McCormack won't let it happen.'

The silence between them gradually coalesced into something almost tangible. Laura had spoken; the subject was closed. There was something in the way Laura always spoke about Gerald McCormack that irritated Angus. She was right, however. If McCormack were against the proposal – and he would be, being in Lord Purdey's pocket – the idea was dead in the water. He would pull strings, manipulate, bully, lie and do everything he could to get his way. Angus looked at Laura and wondered. Maybe there was a way.

***

Monday morning arrived as relentlessly as Monday always does and Angus was sitting at his desk in the open-plan office that he shared with six others in the council's regional buildings. The room was a sterile grey. Before Members' Services had taken it over, the office had been occupied by engineers; their pictures of notable Highland bridges still adorned the walls. There was the Skye Bridge, the Kylesku Bridge, and picture of the old Victorian suspension bridge that spanned the Ness less than half a mile from where Angus sat.

He opened up his computer and began searching for the lynx application that was soon to be discussed in council. He scrolled through the endless planning applications, appeals and other documents until he found it: 'Application for Licence for Rewilding'. The application was long and well researched and he was disappointed to read that the council was only the first barrier to approval. Even if Highland councillors approved the application, the Scottish Government would get the final say. It took an hour of careful reading for him to get through the application. The report referred to similar projects in Europe and went on to talk about the economic benefits in terms of environmental tourism, which could bring a significant boost to the Highland economy.

Then he turned to the objections and found that section empty. That struck Angus as odd at first until he realised that the objections were probably sitting in a filing cabinet somewhere waiting to be scanned. He searched the online system and the name of a clerk appeared, Vicky MacDonald; it was her responsibility to scan the documents. He knew almost everyone in the council and Vicky was no exception. She was a bright girl but more interested in talking with her friends than her job.

Angus waited until lunchtime and then strolled down the corridor just in time to see Vicky and a couple of friends heading off to the canteen. Angus had always played by the rules – in fact he'd spent much of his life making sure that everyone else did – but somehow he felt this was different. He'd felt his perspective change and broaden over recent weeks. Somehow he wanted to make a difference.

Vicky's office was empty when he walked in. He found her filing cabinet unlocked and after a couple of minutes he retrieved the grey folder he was looking for. As he pulled it from the cabinet, he realised that he was sweating and his hands were shaking. He slipped the folder into his jacket and strolled out of the room trying to look as casual as possible. Walking down the corridor back to his office, he realised something had changed. He had changed – he'd become a criminal.

***

Laura watched Angus fork the last few morsels of potted beef into his mouth and force himself to swallow it.

'Delicious, dear,' he said in a monotone.

As usual her husband had no appreciation for the time she had spent cooking the meal. It had taken her hours watching the meat simmering, checking on it every few minutes.

'The butcher said he didn't know the last time any of his customers had cooked that dish.'

'Amazing.'

Angus gulped down a glass of water. Perhaps she should give him beans on toast every night – he probably wouldn't notice.

It's Thursday, she reminded herself. *He'll be going to his precious mountaineering club again, thank God.*

'Thursday night again.'

'Aye.'

'I suppose you'll be going to the club,' Laura said, trying to sound casual.

Angus sighed, as though it would be something he'd hate. 'Well, I suppose I could miss it just for once.'

Laura looked up. 'You got yourself elected president, you have to be there.'

Angus had fallen into her trap: she had made him feel guilty for going out and leaving her alone that evening, and now she would make him ashamed if he didn't go. Either way she would have proved that she was morally superior and everything was his fault.

Angus looked bewildered. 'Well, I suppose … '

She rose from the table and took his plate. 'No, no. You just go, I'll be fine.' She managed to get just the right hint of martyrdom into the comment.

'I'll be back at the usual time,' Angus called as he closed the front door and stepped out into the warm air of the July evening.

As soon as she heard the door close, Laura got off the sofa and walked to the window. She watched, her face pressed against the rough surface of the net curtains, as Angus turned the corner and headed away down the street, disappearing into the darkness where the street lights ended. Once she was certain he was gone, she returned to the sofa and picked up her phone. With deft fingers she dialled a number and a male voice answered.

'He's away. I'll open the back door.'

She deleted the call from the phone, got up and walked into the kitchen, singing softly to herself. Then she unlocked the door and picked up a bottle of red wine from the rack on the work surface and took two glasses from the cupboard.

***

The Thistle Inn was unusually busy that evening as Rory watched Angus steering his way back from the bar with two pints of beer. Rory was sitting at a small table with Brian – who was making himself invisible as usual – and Jen, who had come straight from a long shift at the hospital. A group of younger club members were sitting around a table a few yards away. One of the older members, a tall blond man in his early forties called Gary, was entertaining the group with one of his stories of being lost on Ben Nevis. Rory had heard that tale at least ten times before. Each time the weather was worse and the position more desperate.

As the glasses were placed on the table in front of him, Rory looked up and carried on the conversation they'd been having before Angus went to the bar. 'I can't believe it.'

Angus sat down beside him. 'It's true. Lord Purdey, as large as life, spouting nonsense on the TV. He said that the RSPB are killing birds of prey to give landowners a bad name. He's even started a campaign.'

'Surely nobody believes that! The RSPB killing birds?' Rory said incredulously.

'It's in the *Daily Mail*.'

Rory laughed. 'Oh well, if it's in the *Daily Mail* it must be true. And there's a campaign?'

'Campaign for landowners or something.'

Brian cut in to the conversation, his voice earnest and deliberate. 'Campaign Against the Persecution of Landowners.' Brian might not be very good at talking but he could certainly listen.

No one expected Brian to say anything so there was a surprised silence when he did. Rory waited a moment until he was sure Brian was not going to break with tradition and add another sentence.

'The Campaign Against the Persecution of Landowners! They can't be serious.' Rory put his head in his hands.

Jen had been listening quietly to the conversation, sipping her orange juice. 'I'm sure he's serious. That don't mean he's not mad as well, mind.'

Angus spoke in a low whisper. 'There's going tae be a Highland Council debate on the reintroduction of lynx. It looks a close-run thing.'

'You can bet Purdey'll be sticking his oar in,' Rory said.

'Aye, and he has powerful friends too, yer ken,' Angus said quietly. 'He's no as big a fool as he looks. I took a wee look at the papers. Looks like Purdey has organised quite a lot of objectors. That could swing it. I never expected to see Sinead on the TV, though. She never said anything about it to us when we met her.'

'That's true. I can't get over Purdey, though.' Rory was beginning to feel his anger rise. 'Nowhere in the world is landownership concentrated in the hands of so few. It's bloody feudal. The Campaign Against the Persecution of Landowners! What he means is if anyone so much as suggests they should obey the law and not kill birds of prey, or stop slaughtering mountain hares, *we're* bloody persecuting *them*.' By now he was shouting, and Jen gently put her hand on his arm to calm him. 'Well, it gets my goat.'

'Don't bring goats into it, love – they'll shoot them as well.' Everyone around the table laughed and that relaxed the atmosphere.

There was a hubbub at the next table and Rory looked up. 'Gary, when are you going to get your finger out and finish these Munros?'

The man had a mop of blond hair and bright blue eyes. He looked over to Rory and grinned. 'I thought you didn't like people who followed lists?'

'If you're dumb enough to follow some list drawn up by a Victorian lord I'm happy to let you do it.'

Angus laughed. 'I finished my Munros years ago. I thought I'd do them again but never got round to it.'

Jen shook her head. 'Plenty of time yet, Angus. Ye could allus start again.'

Rory sighed. 'I don't understand it. Over a hundred years ago some bloke with nothing better to do sets about counting all the Scottish hills over 3,000 feet and finds there are … ' Rory searched his memory for the number.

'Two hundred and eighty-two,' Gary called over, then came and sat at the table with Rory and the others. 'I thought you didn't count the Munros you climbed?'

'That's right, I'll leave that to you lemmings. I climb hills not Munros,' Rory teased gently.

Angus had spent a large part of his life completing various lists and wasn't about to let Rory off with decrying his favourite pursuit. 'It's quite an achievement, climbing all the Munros. Took me to a lot of places I'd no have seen otherwise.'

'Boring hills, you mean.'

'Ignore him,' Angus said to Gary. 'How many have you got left?'

'Just the one,' Gary replied proudly.

This got the attention of all the room. A last Munro is always a cause for celebration. Even Rory was secretly impressed. 'Which one is it?'

'Ben Bhuidhe,' Gary announced solemnly. 'I thought we could spend the night in Glen bothy after doing the hill. Have a bit of a do.'

This was clearly a moment for Angus to assume his presidential mantle. 'Do you have a date yet? When yer do I'll announce it in the newsletter.'

'That's right,' Rory said, unable to suppress his grin. 'The headline would read, "Man Wastes Life Completing Pointless List of Mountains".'

'Oh, bugger off,' Angus replied, as the small group of hill-goers collapsed into laughter.

CHAPTER 7

The *Maid of Morvan* was crowded on this June Friday evening. Its crew, assorted men and women dressed in orange hi-vis jackets and clunking sea boots, cheerfully waved the waiting cars on board. Rory was always amazed at how many vehicles the crew managed to wedge on the open deck of the ferry, which was no more than a hundred feet long with a boarding ramp at each end. There were retired couples in their enormous camper vans, bought with their pensions' lumps sums and then taken on to the narrow roads of the Isle of Morvan where they would be driven at a snail's pace. There were farmers with huge four-by-fours towing trailers full of bleating sheep. There were vans driven by plumbers, electricians and telecommunication technicians. The rest were tourists, walkers and cyclists, and families returning to the island to see elderly relatives. Like all ferries heading for the small islands on the west coast of Scotland, it held the pulse of the community. Everything and everyone who came on and off the island passed across its steel ramps.

Sometimes, when gales came in from the west, the ferry could not sail, and then Morvan would become a true island, like a castle under siege, its drawbridge up. At times like that, when white-topped waves filled the Minch, the few hundred folk who lived on the island bowed to nature, with no choice but to let the elements run their course.

Rory stepped out of the Land Rover to stretch his legs on the short crossing. He leaned on the rail and watched the sea churning beneath him as the ferry began its journey. A warm breeze touched his face and the air was full of the cries of gulls circling the boat, hoping to catch the corner of a tourist's sandwich tossed into the air. He could feel the throb of the engines pushing the boat out against the current. To him there was something special in taking a trip to an island. As he sailed across the water he left behind the sterile world of his nine-to-five existence in BetterLife, going to a place where nature was still not quite tamed, where the air had a different smell and he could sense freedom.

He felt an arm slide into his and turned to find Jen smiling up at him, the wind turning her blond hair into a chaotic tangle.

Jen glanced at the view and then back at Rory. 'Nah then, what tha thinking?'

Rory realised he'd been lost in thought. 'I was wondering what it would be like to sail over to an island like this and be going home. Some place with a couple of acres of garden.'

Jen smiled. 'Aye, 'appen we will one day.'

Rory wondered how long that would take, how many more hours he would have to spend caged in that windowless room calculating endless figures for BetterLife or walking down the right-hand side of the corridor with his coffee neatly encased with a regulation lid. They had some savings but not nearly enough to get a deposit for the type of place they wanted. The rent for their town centre flat was a financial sinkhole. Rory watched the mainland receding as Morvan grew closer, its green hills emerging from the summer haze, and felt himself unwinding from the drudgery of his week's work.

Rory looked at the vehicles crammed on to the deck of the ferry. Most weekends there would be no one else from the mountaineering club, just him and Angus in the battered Land Rover. Sometimes, especially in the winter, the only other vehicle might be Brian's old blue Nissan. This weekend it was different. Gary's last Munro had brought a dozen or so club members out of the woodwork. Jen had come with Rory and Angus – for once her shift patterns had given her the weekend off. Gary had two carloads of mates to accompany him up the hill, and a handful of older club members planned to make the trek to the top of Ben Bhuidhe and celebrate later in the bothy. So many had come that a small village of tents would be needed for all of them.

That June evening was hotter than most. The air was thick, warm and sticky as the caravan of backpackers shuffled across the Purdey estate. The old ghillie leaned against a gate smoking one of his evil roll-ups as the walkers passed by, weighed down by their heavy packs full of luxuries for the celebration of Gary's final Munro. As Gary passed, he called out a cheerful 'hello' to the tweed-clad figure propping up the gate. Donald returned his greeting with one of the stares Clint Eastwood used before he gunned down bad guys. No one spoke to him after that.

The heat ate into them as they climbed, shirts stuck to their backs with sweat, and packs pressed down on them. Then they crossed the bridge into the Muir estate – rebuilt already after the flood – and the landscape changed, the air alive with insects living out their short lives in these fleeting summer days. Now the yellow, lifeless grass of winter and the deeper browns of last year's heather were gone, replaced by a riot of green as the earth of the hills burst forth with life.

Rory stopped and turned to Jen. 'You hear that?'

Jen stopped and listened. 'What?'

Rory could hear a bird call, like a series of rising tweets. 'I can hear a wood warbler.'

Jen listened again. 'Oh yes, I hear it too.'

Angus wiped the sweat from beneath his sun hat. 'What's that?'

Rory already had his binoculars up and was scanning the oak trees around them. 'Be hard to see in this … ' Then his eyes caught a small yellow and greenish bird hunting for insects in the branches. 'There he is.'

He gave the glasses to Jen. 'See that broken branch, right at the end of it.'

Jen grinned. 'Ah, I've got him nah.'

'It's a wood warbler, Angus,' Rory said. 'Listen.'

Angus listened. 'I'm nae sure what to listen for.'

'That sort of "peep, peep, peep".' Rory pointed to the tree. 'Give Angus a shot on the glasses, Jen.'

Angus peered through the glasses. 'Ah! Aye, ken him fine now. Are they rare?'

Rory shook his head. 'No, pretty common really. Hard to spot though, always in the trees.'

Angus put down the glasses. 'He's gone. I'll have to get a pair of these.'

Jen laughed. 'You're becoming a twitcher, Angus. Let's get moving. At least there's no midges.'

Rory took a swig from his water bottle and handed it to Jen.

'Too much sun for them.'

As he spoke he felt a sharp pain in his leg and looked down to find a horsefly hungrily burying its incisors in his skin. Rory slapped the little fly and it fell to the ground, dazed. Seconds later Jen and Angus were bitten too.

The hordes of biting insects drove them up the glen and by the time they reached the bothy their arms and legs were pockmarked by bites. The thick stone walls of the old house meant that, even in the hottest days, the inside remained cool. The darkness of the bothy came as a relief from the sun.

'I hate summer,' Angus announced. 'The sun comes oot and everyone says how marvellous it is that they can sit out in the garden and bask in the sun. Hot summers are all right if that's all you want to do. I'm exhausted.'

Rory flopped down on the wooden bench next to him. 'I'll be glad when this heatwave is over.'

Jen was pulling clothes and food out of her pack. 'Listen to you pair of grumpy old men. We don't get many summers like this.'

'Thank God.' Rory pressed the cool water mug against his face.

***

That night the bothy was crowded with bodies. Rory lit a small fire. The night was mild enough not to need its warmth, but Rory thought that a few sticks burning in the fireplace gave the rough dwelling a heart. He was sitting watching the flames take hold when Gary came over to join him. Gary was one of the few native Invernessians in the club. He was a cheerful man, powerfully built, who had played rugby before he became involved in the hills. Gary was in good spirits when he sat down beside Rory.

'How long has it taken you?' Rory asked, sipping his can of beer.

Gary thought for a moment. 'Hmm, let's see … five years, I guess. It's been a kind of quest.'

Rory couldn't resist teasing him. 'Five years of plodding around on rain-soaked summits. You could have achieved something in all that time – become an architect or written a book.'

'To be honest I was never much good at drawing. And surely only an idiot would spend his time sitting in front of a computer writing a book?'

'So what will you do with your life after the Munros?'

'Well, I'm not going up any bloody mountains, I can tell you.'

Angus was sitting with his pipe. 'I bet you'll be doin' another list before you know it. Maybe the Corbetts.'

Rory glanced out of the window and realised that the light was fading outside. 'Look, it's bat time, Angus. Want to come and see?'

Angus nodded and they headed for the door. Rory was going to ask Jen, but she was deep in conversation with some of Gary's friends and he decided not to disturb her.

Outside, the sun was going down, the outlines of the hills softening as day gave way to night. Rory liked being in the hills at night – it was one of the things he liked most about visiting bothies. There was a different side to the mountains at dusk, when the sharp-edged day gave way to the gentle tones of evening. If you spent your days walking in the hills and went home before darkness fell this was a side to the mountains you would never see.

Angus puffed at his pipe. 'It just seems so peaceful.' At that moment there was an outburst of raucous laughter from the bothy door.

'Are you sure about that?' Rory said, and they both laughed. 'I know what you mean. Makes the days I spend in BetterLife feel a long way away. No horseflies now either.'

They watched as the light faded, mountains eased down into the night and the shadows on the hills fell towards the floor of the glen. Darkness crept across the face of the landscape like a warm blanket. Then the first black shape darted around the corner of the bothy, flying with incredible speed as it gyrated effortlessly like a tiny black fighter plane. Soon a dozen or more dark shapes were weaving intricate paths around the bothy roof and away into the trees beyond.

Rory never tired of watching them. 'Pipistrelle bats. It's amazing they can catch something as tiny as a midge by echo-location.'

Angus blew out a ring of smoke. 'Anything that eats midges is a friend of mine, yer ken.'

They fell into silence, watching the bats, and then Angus spoke, more seriously this time. 'I never ken there was so much to see till youse showed me. I've spent years walking in these hills and I've been awful blind.'

Rory had never heard Angus speak that way – usually he was so confident, full of knowledge about the hills. He shrugged. 'The worst of it is, once you know what these mountains should be like and what's happened to them, you just see scars everywhere. Hills stripped bare of trees, hydro schemes and wind farms and bloody grouse moors. It gets hard to look at these mountains without thinking what a mess they are.'

Angus knocked out his pipe on the bothy wall. 'Aye, well, maybe it's time we did something aboot it.'

Rory watched Angus walk back into the bothy. Angus seemed different, somehow, to the man he'd known for the last few years – there was an anger about him that Rory had not seen before. Rory was about to follow him when Jen emerged.

She yawned and stretched. 'I'm knackered. Let's get into our tent.'

Before Rory could protest Jen took him by the arm and led him to their tent. Normally he would have slept in the bothy, but with so many folk about he'd carried in the tent – space indoors was limited and likely to be noisy, and it gave them a place of their own.

Rory woke to the sound of a gentle breeze tugging at the nylon flysheet. It was still early but already the warmth of the sun was making it uncomfortably hot inside the tent. He lay still for a while with Jen nestled close to him, still sleeping. He could feel the rise and fall of her chest against his back and the touch of her breath on his neck. It felt warm and comfortable and it was nice having her so close to him.

An hour later, the clatter of pots and pans filled the bothy as the whole group tried to brew up tea and coffee, cook porridge, or fry bacon. Outside, the sky was blue with the odd white cloud drifting by. Rory was making tea and stirring his porridge when he saw Gary wandering over, waving a delicious-smelling bacon sandwich dripping with grease and oozing tomato sauce.

'Tell me you wouldn't fancy a bacon sandwich, Mr Vegan.'

Rory laughed and pushed it away even though his mouth was watering. 'Get thee behind me, Satan.'

'I won't tell Jen. She'll never know.'

'Charred corpse, that's what that is,' Rory snorted.

Gary crammed the roll into his mouth and took a huge bite. 'I know, and it's lovely. Mmm.'

Rory looked around the bothy. Angus was trying to organise everyone and failing miserably as usual; Brian was nowhere to be seen. 'Is Brian away?'

Gary had swallowed the last of his bacon roll and was wiping tomato sauce from his cheek. 'You know Brian. Always up with the larks. Said he might see us at the top of Ben Bhuidhe but you never know with him.'

Jen was sitting in the sun on an old wooden bench against the bothy wall. Rory handed her a mug of tea and a bowl of porridge peppered with blueberries. They ate silently, looking out over the forest to the hills beyond. Rory watched a house martin coming in from hunting insects in the air, its black wings swept back and white underbelly showing against the blue sky. He watched where it landed in the little cup-shaped nest it had built beneath the eaves of the bothy. No sooner had it fed its chicks than it was airborne again, seeking more sustenance for hungry mouths. Moments later it was back, bobbing through the air, beak crammed with insects.

Jen had been watching it too. 'Hell fire. Them chicks keep her busy.'

'All the way from Africa to nest on this bothy wall. It's amazing.'

Jen watched the little martin take off again. 'You wonder how it could come all that way to make a house right 'ere.'

'Won't it be great to live somewhere like this? We'll get our own place somewhere like this one day.'

'Aye, maybe.'

Something in the way Jen spoke made Rory feel that she was a lot less keen than he was to move to the west coast. 'You do want to go, don't you?'

Jen stared at the grass. 'Aye ... but maybe not yet. I'm doing all right at the hospital, and ... well, let's just see.'

Rory felt his stomach churn. Perhaps Jen wasn't as committed to moving away from Inverness as he thought. 'Do you not want to go?'

Jen shrugged. 'Aye, but 'appen not yet.'

Rory got to his feet, his head filling with thoughts of years and years in the enclosed cell of BetterLife. 'I can't stay in that bloody factory much longer. I just can't fucking do it.' Rory stormed off, and Jen let him go, sitting silently against the bothy wall.

***

The gaggle of walkers climbed higher, and Angus watched as the surrounding hills seemed to come alive in the shimmering heat haze. The ascent of the hill was a long, sweating plod, the gradient unrelenting and the horseflies just as persistent. At last, close to the summit, a gentle breeze rose up, cooling the midday heat and grounding the persistent flies.

All the way up, Angus had noticed that Rory and Jen were keeping apart, and he was astute enough to realise that something had caused tension between them. They always seemed so close and happy together. Last Munro completions – or 'compleations', as they are known by tradition – are always jolly events, with fancy dress and music often accompanying the walkers. Gary's was no exception. He was wearing a makeshift trilby hat with the number '282' attached to the top, but the figures looked as if they had been stolen from birthday cakes and still had icing on them that was melting and running down the hat. Gary smiled and chatted to his friends while someone beat an irregular rhythm on a drum.

At the top of the hill one of Gary's friends produced a set of bagpipes and managed a ceremonial serenade as Gary took the final steps to the cairn. The tune might have been *Top of the World*, although it could well have been something else – it was fairly hard to say.

Angus opened his rucksack and pulled out a flat cap and a scroll. He made a show of offering the cap to Gary, who replaced the trilby with it. Club tradition dictated that the Munroist should wear the ancient flat cap – one worn, or so legend has it, by the famous climber Don Whillans. Gary knew what was coming and, once enthroned on the summit cairn, waited for Angus to read.

Angus stepped forward and read from the old battered scroll.

> '*He has done the lot,*
> *This hill-climbing clot.*
> *And though his feet are sore,*
> *At last there are no more.*
> *So now he's reached this top,*
> *He's going to have to stop.*
> *So all you gathered here must surely know,*
> *He's climbed his last Munro.*'

With that Angus tipped a glass of whisky over Gary's head and the assembled company let out a loud cheer. Then the whole club gathered and indulged in a great round of recording the moment in photographs. In the crush at the top Angus noticed Jen take Rory's hand and squeeze it. He was pleased to see them together; any fool could see the love between them.

Angus looked around the group but couldn't see Brian. He called to Rory over the cheering. 'Nae Brian, then?'

Rory got himself higher on the cairn, checked the folk on top of the hill, then scanned the horizon. 'There's someone over there, on that ridge.'

Angus could just make out a figure perhaps two miles away heading down a grassy ridge.

Rory took out his binoculars. 'It could be him – I can see a blue top just like that scruffy one he wears – but he's a hell of a way off.'

Angus shrugged; Brian was always a law unto himself. 'Aye well, we'll see him in the bothy after the walk.'

Jen took Rory's hand and they set off down the hill, with Gary and his friends singing songs and joking in the warm summer afternoon. Angus felt content as they walked. If they could all be as merry as this on a summer's day in the hills perhaps there wasn't too much wrong with the world after all.

'Now then, what's tha going to do now there's no more Munros, Gary?' Jen teased as they headed down the mountain back to the bothy.

Angus cut in. 'There's always the Corbetts, yer ken.'

Rory shook his head in despair. 'Oh God, is there no end to it?'

They got back to the bothy tired but in good spirits. The plan was to have a barbecue to celebrate Gary's achievements. There is no greater accolade to the human spirit of optimism than to plan a barbecue in the Highland hills. There is more chance of downhill skiing in the Sahara. If it is not far too windy – an all-too-common occurrence – then the midges will take advantage of the light winds and devour everyone. If it is not windy and there are no midges then that can mean only one thing: it's raining. In the Highlands all three of these pestilential curses can strike in less than an hour.

Angus sprayed himself with repellent but still the tiny creatures came like miniature kamikaze planes thundering into the deck of an aircraft carrier. They attacked everywhere his skin was exposed, even buzzing into his ears and invading his nostrils. He wondered how long he could stand to be outdoors.

Rory was setting up the barbecue, pausing every few seconds to swat away the midges tormenting his skin, when he looked up at Angus. 'Sod this.'

Gary was unwrapping sausages. 'Hmm, I think you're right – let's go in.'

Inside the bothy they were safe from the marauding midges. Angus lit his pipe and set his tin of Irish Cask open on his lap, relieved to be away from the tormenting insects. Rory set the barbecue up in the old hearth and coaxed flames from it.

Soon there were plates of sausages and burgers being passed around, along with tins of beer and drams of whisky poured into plastic mugs. Conversations varied from the profound to the profane but all followed the old bothy tradition of exchanging craic well into the night. Bothy craic is a type of banter that mostly consists of abuse and insults. The more severe the insults, the greater the affection amongst the company.

Jen handed Rory a plate laden with bread, couscous, hummus and salad. 'There – fill tha boots.'

Rory and Gary sat together on a bench eating their meal, Gary with his meat and Rory with his vegan food. Angus sat quietly beside them. The food, the whisky and a good day on the hill were all making him sleepy.

Gary finished and put down his plate. 'I'm full. No more for me.'

Rory smiled. 'Was that four burgers you had? Still, I suppose 282 is a hell of a lot of mountains to climb. You need a feed after that.'

Gary opened another can of beer and took a sip. 'Now that it's all done, I'm not sure why I did it. People at work ask me, they say, "Why bother going up all those hills, Gary?"'

Rory was suddenly serious. 'It's why *wouldn't* you do it that bothers me. Why wouldn't you want to wander these hills, be out in the wind and see the sun come over a mountain ridge? This is where we came from. We evolved to wander the landscape, hunting and gathering food. We were born to explore, born to be curious, born to wonder what's over the next ridge, what it's like on that summit. It's why we're here.'

'I never thought of it like that.'

Rory fumbled with his plate and fork as he grew more animated. 'We should wonder why we do anything else. What's odd is hours wandering round the supermarket or lying on the couch glued to a TV screen, or locked in some factory while another day of your life ticks by. Ask them why they do that – that's the real question.'

Angus nodded. He understood why Rory felt that way. The need he felt to wander the hills had been something visceral for him all his life. Perhaps his recent doubts were nothing but the pangs of getting older.

Around nine in the evening the bothy door creaked open and Sinead, the Muir estate manager, and her assistant Hector joined the group. Sinead carried some cans of beer and a bottle of wine along with a bag that had French bread sprouting from it. Angus noted with concern that Hector was carrying his guitar, although he needn't have worried – as soon as Hector began playing everyone sang along so heartily that the vagaries of Hector's musicianship were of no consequence.

Sinead watched Hector playing for a few moments before she waved to Angus and Rory and brought them into a con-spiratorial huddle. Angus thought he'd never seen her look so serious.

Angus welcomed Sinead with a smile. 'I didnae ken yer were coming tonight.'

Sinead shrugged. 'Now I wouldn't miss a hooley, would I?'

Then she lowered her voice. 'There's something youse ought to know.' Her eyes shone bright in the gloom of the bothy. 'We had a visit from Lord Purdey's lawyer, Gerald McCormack. He's a little-blood sucking leech, so he is.'

'The councillor?' Angus drew closer when Sinead nodded, sensing that this was something he should know about. 'What did he say?'

'You know how this bit o' land we are standing on now was

lost by that eejit Lord Purdey in a card game in the 1970s?'

Angus and Rory nodded. Like everyone else, they'd heard the tale – it was a local legend.

'Well, they sealed the deal on a napkin, but it got lost a long time ago. And thon hallion McCormack says if we can't find it then Purdey gets the land back.'

'Because wi'out the deeds you cannae prove it's yours,' Angus said. There was a hollow sinking feeling in the pit of his stomach as he imagined what Purdey would do to the land if he got his claws into it.

'Tony's sketchy about what happened to the napkin,' Sinead said. 'I've been taking the office apart. It could even be here.' She glanced around the bothy and took a sip of red wine.

Angus's mind whirred with possibilities – most of them unpleasant.

'And I'll tell youse this,' Sinead added quietly, 'first thing Purdey will do is close this bothy. Probably demolish it or turn it into a holiday cottage.'

'Aye,' Rory said, his voice tight. 'And he'll take down this forest and turn it into grouse moor.'

Sinead put a hand on his shoulder. 'I know you love coming here.'

Rory was suddenly angry. 'It's not right. Just because he gets hold of a piece of paper some rich man can decide what happens to a place like this.'

Angus sighed. 'Well, he does own it. It's up to him.'

Rory struggled to find words at first but then became eloquent with rage. 'These are our hills – this bothy and these mountains belong to everyone. People like Purdey ... they got *rights* over these places, but they don't *own* them like you can own a TV or a pair of trousers. Land is different.'

Angus spent all his life dealing with council rules and regulations. They had become his working life and he couldn't see how they could be ignored. 'Ownership is nine tenths of the law.'

'Bollocks!' Rory threw down his beer can. 'Purdey didn't make this land; he just took it. Maybe three, four hundred years ago, one of his ancestors turned up with a load of blokes with big sticks and told everyone the land was his. And since he had more blokes with sticks than anyone else he kept it.'

'But that's the law,' Angus managed to say as Rory drew breath.

'Bugger the law. Who do you think made the law? Land is power.' Rory fell silent, but Angus could see that the anger still burned in him.

Sinead shifted uncomfortably, as if guilty that she had ruined the atmosphere of the evening. 'Maybe we'll find them deeds. Who knows?'

In the far corner of the room Hector toppled off his chair, to the cheers of those around him. His guitar fell silent and someone began to recite a poem. At last the fire died, the songs ceased and even the hard core of revellers gave in to their sleeping bags in the bothy or wandered out to their tents.

***

It was around seven o'clock in the morning when Rory woke, his full bladder summoning him from a dreamless sleep. He rolled over and tried to seek the comfort of that dark oblivion once more, but a full bladder is a restless thing and eventually he was forced to crawl out of his sleeping bag and unzip the tent door. The day was young and the early morning cold, with dew heavy on the grass. In the still air, fingers of daylight were beginning to stroke the face of the mist that hung in the glen, shrouding the forest in grey curtains. As he peed, a deer and a fawn stepped out into the clearing near the bothy and were, for a few seconds, heedless of him; then the doe caught his scent, barked in alarm, and the pair vanished into the forest.

He decided to head into the bothy in search of tea. It took his eyes a few seconds to adjust to the gloom, but he soon spotted Angus sitting waiting for the kettle to boil, filling the bothy with the hiss of his stove.

'All right, Angus.' Rory yawned and stretched.

Angus lifted the lid of the tin kettle and steam poured out into the room. 'Aye, I'm a bit tired. We were up late last nicht, yer ken. Tea?'

Rory cast about the bothy. The floor was a mixture of beer cans, half-drunk mugs of wine and sleeping walkers. Then he noticed Brian's sleeping bag rolled up against a wall.

Rory took the mug from Angus. The tea was hot, and he was grateful for it. 'Brian's away already, I see.'

Angus was still sleepy as he sugared his tea. 'Aye, that's the way of it.'

Rory glanced at the sleeping bag again and a chill went down his spine. He couldn't remember seeing Brian at the party last night. Then he told himself that it had been crowded and Brian was quiet. He must have been there, unnoticed in the hubbub. Rory searched his memory again, this time with growing panic.

He tiptoed through the chaos on the bothy floor and felt the sleeping bag with the palm of his hand. It was cold. Rory felt his stomach turn in knots.

'Angus,' he said slowly, 'did you see Brian this morning?'

Angus was occupied with spooning porridge into his pan. 'Nah, must have been away early, yer ken.'

Rory stared at the cold, empty bag. He was wide awake now, running a video of last night through his mind. No matter how many times he did that, Brian didn't appear in the picture.

Rory picked his way back to where Angus was standing with his back to him, making his breakfast, and asked the question slowly, afraid of the expected answer. 'Did you see him last night?'

Angus froze, then dropped his spoon, which scattered a cloud of porridge as it fell. He turned, grey-faced. 'Nah, I dinnae ken. I never did!'

The next few minutes were a chaos of shaking folk awake in their sleeping bags. They ran from one to the other in the hope that someone had noticed Brian the previous night, but the bleary-eyed revellers just stared back in confusion. Soon they'd spoken to everyone. No one could remember seeing Brian.

Angus held his forehead in his hands, trying to squeeze the memory from his brain. 'He nivver came back.' Angus was almost crying. 'I should have realised – why did I nae realise?'

Then Rory was running up the hill behind the bothy, gasping for breath, moving as fast as he could. *Higher, higher*, was all he could think. Minutes, even seconds seemed important now. He paused, glanced at his phone. *No signal. Higher!*

He ran on up the hill, his legs screaming with the effort.

Glancing down, he could see figures moving below him in the glen, people scouring the paths and the edges of the forest. Their cries drifted up to him. 'Brian! Brian!'

At last, on the fourth stop, he looked at his phone and it showed two bars. Chest heaving, gasping for breath, he tapped 999 into the phone.

The calm, gentle voice of a young woman answered. 'Which service do you require?'

Rory fought to control his breathing. He hadn't thought what he was going to say. 'Mountain rescue, someone's lost,' he gasped between breaths.

'I'll put you through to the police then, sir,' the young woman said, as if she were booking him a table for four on Friday night at the Indian Ocean on Academy Street in Inverness.

Moments later he was talking to a desk sergeant. 'I see, and when did you last see this gentleman?'

'Yesterday,' Rory answered, but then he realised that wasn't true. 'No, no, wait … it must have been Friday night.'

The policeman's tone became more serious, almost accusatory. 'So he's been missing two days, then?'

Rory felt even more confused. 'No, no. He left early morning on Saturday.'

'What time?'

'Er, I don't know.' Rory realised that he was sounding more and more hapless. He should have thought through what he was going to say.

The police sergeant was relentless, methodical, almost mechanical. 'What mountain would that be, sir?'

Rory mumbled something in his best Merseyside Gaelic and

the policeman grunted.

'Can you spell that?'

No, he couldn't. The inquisition continued. He wanted Brian's date of birth, then his address, which Rory remembered by some miracle. Then he wanted Rory's name and address – which, mercifully, he recalled without effort. At last the police officer was content he had all the details he was going to get from the confused and panicking hiker at the other end of the phone.

There was a click, and then he was speaking to Chris, the mountain rescue leader. This was someone he could deal with – one of his own who spoke his language.

'Ah yes, Glen bothy, I know it. You'll have been up Ben Bhuidhe, then?' came the clipped military tones of the team leader. 'OK, keep the rest of your party together. On our way.'

An hour later a small, forlorn group stood outside the bothy in the early morning light. They spoke in hushed whispers and felt as if a great weight pressed down on them. Minutes later there was a distant rumble which grew until the glen filled with the roar of an engine as the red-and-white helicopter swung into view.

Rory watched the aircraft turn into the glen and swing towards the bothy. He tried to fight it, but a sickening sense of dread slowly filled his mind.

***

Angus turned his head away as the downwash from the helicopter hurled leaves and sticks into the air. It felt as if the whole glen was shaking as the aircraft came in to land. A tall, slim man with sharp chiselled features jumped down first, followed by a well-built man. They ran towards Angus and Rory, keeping low to avoid the whirling rotors. As soon as they were safely on the ground the machine began to roar again; in seconds it was hundreds of feet above them, and turned away across the glen to search the hillside.

The tall man shook Angus's hand. 'I'm Chris, the team leader, and this is Iain, my deputy.'

Iain – who sported a beard and a blond ponytail and reminded Angus of a Viking – smiled, and they went into the bothy. It was cooler inside and Angus spread his map out over the table.

Rory pointed to a ridge on the map. 'I think we last saw him here, maybe quarter past two.'

Chris put up his hands. 'Slow down now, hey. Tell us a bit about Brian. How experienced is he?'

Angus put his fingers together as though praying. 'Brian's about the most experienced walker I know, yer ken. He's very keen, always walks alone, and he's up and away before all of us. But he's quiet like. Yer nivver ken what's in his heid.'

Rory could tell that Angus's Doric had confused the team leader. 'He's a bit of a loner, Brian, quiet bloke,' he translated. 'You never quite know what's going on in his head.'

Chris looked thoughtful. 'I see. Did he tell anyone what his plans were?'

Jen answered him. 'Brian allus kept his sen to his sen. I don't think he even knew his plans till he set out the door.'

Angus noticed she had her hill boots on and was ready to head out. 'Are you going looking for him?'

Jen was tying back her blond hair. She looked businesslike. 'Tha don't think I'm going to sit all day on my arse doing nowt wi' Brian missing, do you?'

The Viking looked up from the map. 'That's kind of you, but better left to us.'

Angus touched Chris's shoulder. He was tortured by the fact that he hadn't noticed Brian's absence. 'We want to help. He's one of ours. I've eighteen men and women, all experienced on the hill.'

The rescue leader steered Angus into a corner away from the rest. Then he spoke gently. 'I understand – it's hard for you. Don't blame yourself. You couldn't have known.'

Angus was close to tears. 'It's my club. I'm the president. I should have spotted he wisnae back.'

Chris was firmer this time, his voice terse and military. 'It makes no difference. If you had called him in last night, we'd have said "wait until morning". We wouldn't even have classed him as missing until now. Sometimes people are just late. You know that.'

Angus nodded, his eyes fixed on the floor, unable to raise his head.

'Look, we will use your chaps. They'll need to sign a form and I'll send a couple of men out with you. How's that?' the leader continued.

Angus looked up and nodded. 'We have to do something.'

He turned to bring the club together but Chris took his arm.

'Make sure your folk are careful – no heroics. I don't want two casualties to search for today.' Then, when Rory was out of earshot, Chris turned to Angus again. 'Can I have a word?

I hate to bring this up. Brian ... you told me he was a bit of a quiet bloke.'

'Aye?'

'We can't know what's happened. Could be a fall. Maybe he's been taken ill or something like that.' Chris's voice trailed off. 'It's only, I have to ask ... '

'Ask me what?'

'Do you think he might have been depressed? We can't rule out suicide.'

The word *suicide* hung dark in the air between them. Angus didn't hear it at first and the look of bewilderment on his face made Chris repeat it.

'I'm sorry. I know it's difficult, but I hear he was a bit of a loner and sometimes those sort of people ... '

Angus digested the words. 'I dinnae think he'd have done that. Brian?'

The last word was a question, but it wasn't for Chris; Angus was asking himself. All that came into his mind in answer was darkness.

The quiet bothy was transformed. Land Rovers pulled up outside, police radios chattered, and search dogs barked and strained, their noses keen to seek out their quarry. A generator chugged away outside; inside there were charts on the walls with pins and areas marked off. Chris and Iain were hunched over a laptop showing a three-dimensional map of the island, marking out search areas and trying to work out alternative routes Brian might follow if he had taken a wrong turn or gone off-route for some reason.

Then Iain gathered the club members together in a clearing.

'I'm the centre of the search line. Always make sure you can clearly see the person next to you. You are responsible for the area between you. Make the search slow and thorough, and try to keep the line together. I'd rather we took a long time and were sure the ground was clear than rushed and missed someone. If one person stops to look at something, we all stop. Is that clear?'

There was a murmur of assent and the line moved forward. It was hard going, searching through bushes and bracken – they weren't walking on paths. They had to plough through gorse and look under bushes. In the shelter of the woodland, the midges made them suffer.

They searched for hours until the light began to fade and their legs dragged. As evening approached the sky became pillowed with clouds and a light drizzle started to fall.

The police had found Brian's old blue Nissan still parked where he'd left it at the start of the track to Glen bothy, so they knew he hadn't gone home. The day had been warm and clear with little wind and cloud – perfect for the helicopter. The pilot followed the well-known routes and paths, the hills laid out below the great whirring machine like a huge three-dimensional model. The helicopter and the rescue teams swept huge areas of the high ground. Down in the forests, the searching was slower. A man could lie behind a bush or be hidden by a fallen tree.

The first night might be a twisted ankle or even a broken leg. In summer it might mean a dark and lonely wait for the dawn. In the first night there is hope, but by the second that hope begins to dwindle. By the third night it has changed to desperation.

At last Iain turned and faced the line. 'OK, that's enough for today. Let's get something to eat and drink.'

Angus rushed over to him, anxious to make the best of the search. 'We cannae stop – we've no found him. We can keep going. Can't we, lads?'

'Folk have to rest, Angus,' Rory called out, face lined with sweat.

Angus was tired too – his back ached from stooping to search under bushes – but he would not give in. 'OK, ten minutes then. We move on in ten minutes.'

Iain shot a look of concern to Rory, but it was Jen who took Angus by the arm and spoke quietly to him using her nurse voice. 'Nah then. Tha come back to the bothy wi' me.'

Jen was nursing him now, Angus realised; he looked at Rory, asking the question with his eyes. Rory nodded. Angus let himself be led towards the bothy. He felt tired suddenly and his legs were wobbly; by the time they reached the bothy Jen was holding him up.

'Let's get thee a brew.'

Angus was standing in the bothy, his eyes adjusting to the light, when he noticed a tall blond young man sitting on the bench smiling up at him, his head a mass of curls. Angus recognised him at once, felt a rush of relief and grinned back.

'Ah, yer here, James! I was afraid … '

James shrugged. 'You got any smokes? I'm gagging.'

Angus fumbled in his jacket, searching for his cigarettes, but there was a tear where his pocket should have been.

'I've ripped my bloody jacket.'

Angus showed James the jagged tear. He reached up to scratch his beard but found his cheek bare – he must have shaved that morning. His hands felt rough and cold against his skin and he realised that he was wearing his woollen mitts, caked in ice. The hut was noisy, full of climbers, karabiners and ice screws jangling at their waists, kettles steaming away on gas rings. Angus noticed his breath misting in the cold air. Someone spoke, but the words were garbled and he couldn't make them out. Angus pulled off the iced gloves. His left hand was twisted and blue, the fingers contorted at strange angles.

James smiled at him, blue eyes sparkling. He looked at Angus's hand and laughed. 'You've broken your hand, you daft bugger.'

Angus look at the crooked fingers and tried to remember how it happened. 'Aye, I must have.' He looked back at James. 'I dinnae ken you'd be here.'

James sniffed. 'I've always been here.'

As James spoke, a tiny spot of blood formed on his nose. At first it was no more than a pinprick, but it grew until a thin crimson line of red trickled down past his lips and dripped from his chin.

Angus took hold of James's collar and gently pulled him closer to get a better look. 'James, you all right?'

'I'm fine. Are you OK?'

'Angus?' the garbled voice said again. 'Angus, it's Chris – are you OK?'

Angus found he wasn't holding James at all; it was Chris's collar he held. He felt stupid. 'I'm sorry, I thought you were someone else.'

Angus turned. Jen was beside him looking up, her face a bundle of worry. 'Angus, is tha all right?'

Angus nodded. 'Aye, fine.'

He stepped away from Chris, turned and headed for the door, but on his second step his legs failed and he found himself staggering. Then the bothy floor leapt up and hit him on the shoulder. He fell on through it, rock and ice flashed past, the rope at his waist went tight, and there was a crack, then silence.

CHAPTER 8

Simon Partington was not a morning person. He slumped at his desk in the *Caledonia News* offices and glanced out at the Moray Firth, the wide stretch of water that separates Inverness from the Northern Highlands, and watched the traffic moving slowly across the huge arch of the suspension bridge. He swallowed a mouthful of coffee and allowed the strong, sweet taste to zigzag down into his stomach. At this time of day caffeine was his saviour. He pulled open his desk drawer and found a Twix sitting there minding its own business. Simon looked at it for a moment, glanced down at his stomach hanging over the top of his trousers, and slammed the drawer shut. He opened his emails instead. There were invitations to talks, notifications of housing association meetings, and – most exciting of all – his invitation to the Highland Horticultural Society's annual heather-growing event.

Simon leaned back in his chair and scrolled through the emails, looking for something remotely interesting, but found bugger all. *Sod it.* He pulled open his desk drawer, grabbed the unsuspecting Twix and tore off its wrapper.

He was savouring the sweet molten indulgence of milk chocolate and toffee when his phone rang. 'Simon Partington,' he said, spitting crumbs out over his keyboard.

There was a cough at the other end of the phone and Simon

was sure he heard a small dog bark. 'Purdey here, old chap.'

It was too early for Simon's mind to function and he struggled to put a face to the name. 'Er, right.'

There was another yap at the end of the phone. 'Will you stop it! Not you. Purdey, Campaign Against the Persecution of Landowners.'

The image of a dachshund attached to the end of his leg flashed through Simon's mind. 'Ah yes, I remember.'

'Sorry about your trousers. You left a card, said to call you.'

A profound sense of gloom fell over Simon as he waited for Purdey to reveal yet another transgression by the RSPB that no one would be interested in.

'You know there's a missing hillwalker in this area right now?' Purdey continued without waiting for an acknowledgement.

Simon pulled the story up on his computer screen and scanned it quickly. 'Brian Murphy, sixty-three ... missing in Morvan ... last seen ... et cetera. Concern increasing.'

Purdey coughed again. To Simon it sounded like the nervous habit of someone who knew he was up to no good. 'I read that, yes. But there's one thing it doesn't say.'

Simon was scrolling down the rest of the news items, waiting for Purdey to finish so he could get on with his day, and was now idly reading a review of a play at the theatre, Eden Court. 'What might that be?'

Purdey coughed twice. 'They were having a shindig – the Highland Mountaineering Club, that is – in the bothy when this chap went missing.'

'I see.' Simon didn't think the headline 'Mountaineering Club Has Party' would grab much attention.

Purdey coughed. 'Well … the thing is, they never spotted this chap hadn't come back until the following day. All four sheets to the wind in the bothy. *Quite* the rave. They should have known he was missing in action that night, shouldn't they? One doesn't like to apportion blame at a time like this, but—'

A light went on in Simon's journalistic brain. 'Rowdy Mountain Club Fails Lost Walker' might just be a good story after all. No doubt his editor would insist on him adding the word 'Fury' somewhere.

'I see. Thanks for the information, Lord Purdey.'

Simon put down the phone and googled 'Highland Mountaineering Club'. He quickly scanned through the club's website. *President: Angus Sutherland …*

***

On Monday morning Angus sat at his desk, palms resting on the cool polished surface of the wood. They had told him to rest, tried to persuade him to take the day off, but he had insisted on coming in. He had minutes to write. After all, nothing had really happened. He'd been tired, perhaps drifted off a bit; now he was back to normal.

He arranged the papers on his desk without reading them. His gaze continually wandered across the slate roofs of the town and out to the green hills beyond. He followed the outline of Ben Wyvis less than twenty miles from where he sat, and it drew his mind out on to the hills. Somewhere, on a different hill, Brian must be lying – perhaps with a broken leg, or even worse. Why hadn't he noticed that Brian hadn't come back? The evening played in a constant loop in his head. He was laughing with Rory, chatting to Jen, teasing Gary, eating sausages, sipping whisky, smoking his pipe. All the while he had been doing those things, Brian's sleeping bag had been sitting rolled up against the wall, a mute testimony to the fact that its owner had not returned. He hadn't seen it – why hadn't he seen it?

'I brought you some tea and a few digestives.'

A voice cut through Angus's thoughts. Jean from Accounts stood in front of his desk with tea and biscuits held in her small chubby hands like an offering to the gods. She smiled. 'You must be worried sick.'

He could take anything but sympathy and suddenly was afraid he might cry. He took the biscuits and tea wordlessly, worried that if he spoke all his fears might come tumbling out.

Jean rubbed his shoulders. 'They'll find him, you'll see.'

The tea occupied his hands long enough for him to get some control back. He stared at the black face of his phone but no matter how much he willed it, it refused to ring.

He spent almost an hour moving papers around but couldn't apply himself. The handful of planning applications seemed so inconsequential and he passed the day replying to the odd email and making a few phone calls. Half a dozen folk dropped by – people who thought they should say something but when the time came they weren't sure what.

After work Angus took the ten-minute walk home through the streets to his quiet semi-detached house. Every few minutes an image came into his mind of Brian lying in a contorted heap on the hillside. He was alone and cold and Angus could do nothing for him. He repeatedly pushed that image away but it kept coming back, like a toothache he couldn't cure.

When he turned the corner into his tree-lined road, there was a large young man with long dark hair talking to Laura on the doorstep. They both turned as he walked up the road and the man smiled. Angus noticed that Laura was grim-faced.

***

Simon flipped open his iPad and put it down on the kitchen table. He typed as he spoke. 'Have you been in the mountaineering club a long time, Mr Sutherland?'

Angus answered him quietly, knowing that such questions were only asked to put the subject at ease. 'Twenty years, maybe.'

Simon's eyes widened. 'You must be very experienced. Was Mr Murphy very experienced?'

Laura stopped stirring the tea. 'You mean *is* Mr Murphy experienced.'

Simon stopped typing and looked embarrassed. 'Oh yes, of course, I'm sorry. Is he?'

Angus nodded. The image of Brian came into his mind and he pushed it away. 'Aye, yer ken, very experienced.'

Simon didn't look up from his tap-tap-tap on the touchscreen. 'And did he leave word where he was going?'

The reporter had paused very slightly before he asked that question. It was only a fraction of a second, but long enough for Angus to know that these were the questions he had come to ask.

Angus swallowed. 'Not exactly. We ken the area but Brian kept himself to himself.'

Simon didn't type out that answer; he'd obviously known it already. 'And when did you realise he was missing?'

'Sunday morning.'

The reporter shifted uncomfortably in his chair. 'After the party, then?'

Laura put a cup down on the draining board so hard that both Angus and the reporter turned to her.

Angus felt the muscles in his back tighten. 'There was a wee hooley. Aye. Club member's last Munro, yer ken.'

Simon did type those words. 'So no one noticed he hadn't got back the previous night?'

Before Angus could answer Laura came back from the sink and stood over the reporter. 'I think you'd better be going.'

Simon smiled up at her. 'Just a couple more questions.'

Angus coughed. Laura had a way of taking charge. She had that kind of belligerent presence only small people are able to summon.

Simon paused and flashed another of his reporter's smiles. 'Mr Sutherland—'

Laura's hand on his shoulder stopped him. 'I think you had better go now.'

***

Jen hated night shifts. Even after five years she couldn't sleep during the day and always ended up slumped on the couch while daytime TV blurred across the screen. Today was even worse. Rory was at work and she'd sat for most of the day with the phone in her hand, just in case. It was five-thirty now; it hadn't rung. He'd be home soon, hungry as always, his eyes sore from staring at his PC. Dehydrated from hours in that hot lab.

She walked the few feet from the lounge into their tiny

kitchen, her bare feet sensing the change as she passed from the carpet to the chill of the laminate floor. A pan of chickpea curry was sitting on the hob. She'd made it how Rory liked it – not spicy enough for her taste but just right for him, even though he'd probably complain that it was too hot. As she stirred it, the spoon rattled against the side of the pan and the little kitchen slowly filled with steam that carried the luxurious aromas of cumin, coriander and turmeric.

She heard Rory's feet echoing in the hallway. Jen could always tell his footsteps on the concrete steps up to their flat – no one came up the stairs quite as fast as he did. She knew him in the way couples who have been together for several years get to know each other. She knew how he held his fork, the little noises he made falling asleep and how he would always lose his keys overnight.

In a moment he'd fumble with the latch, enter the flat, and shout, 'Hi, it's me.'

And Jen would laugh. 'Aye, I thought it were thee,' she'd usually say.

Usually he'd come into the kitchen and kiss her in greeting, but today he passed by the kitchen door in silence and slumped on their old fake-leather settee.

Jen followed him into the lounge. 'Hear owt?'

Rory sighed and shook his head. 'Nobody said anything. Nobody asked me if Brian was OK, if he'd been found. Nothing.'

She thought he looked more tired than usual. Maybe thinner too. 'I made your special curry.'

Now, Jen thought, *he'll say 'oh great' or 'lovely', like he always does.* But he ignored her.

Rory scratched his beard and rubbed at his eyes with the backs of his hands. 'Bastards. Not one of them bothered to ask anything. Like I don't exist.'

Jen lay down beside him and nuzzled his shoulder. His arm dropped around her, but she knew he wasn't really with her; his mind was somewhere else. 'What kind of place is that? Nobody even noticed me.'

Jen took hold of his wrist and pulled him in to her. 'Did tha call him?'

Rory picked up his phone and stared at its silent screen. 'Angus? No, thought I better not.'

At that moment the phone came to life. After a start, Rory glanced at the screen. 'It's Angus.'

He put the phone to his ear. Jen watched his face get the pinched expression he always had when he was worried.

Rory listened for a moment and then whispered to Jen. 'No news; search is off for tonight. Angus, Jen's here. Let me put you on speakerphone.'

'They'll search again the morn.' Angus's voice crackled over the phone and Jen thought he sounded strained.

'Oh, right.' Rory sighed and stared at his phone.

Jen realised that she knew very little about Brian. 'What about his family?' she said, raising her voice slightly in the way people did when using speakerphone.

Angus's disembodied voice came over the phone again. 'He's a son down south. The police traced him, yer ken. Dinnae think they kept in touch much. Brian was married once. Never mentioned her though.'

'I guess he never said much about anything,' Rory said.

Angus sounded calm over the phone. 'Maybe we never asked him. He was quiet and we jis' left him.'

'You don't think ... ' Rory stammered. 'What mountain rescue said ... '

Angus sounded calm, but every now and again Jen thought she heard his voice break a little. 'I dinnae ken what to think. We can jis' hope for the best.'

After Angus ended the call, Rory and Jen sat for a few minutes listening to the sounds in the street. Then Jen took Rory's hand. 'Let's have us tea, pet.'

Rory reached for the remote and switched on the TV. 'Maybe in a bit.'

Jen knew he was really anxious then; Rory was always starving.

The news came on the TV. There was a feature about a rocket launching station that was to be built in a remote part of the Highlands.

Rory hissed at that. 'There's another wild place gone. Soon there'll be nothing left.'

Then the TV showed a red-and-white helicopter taking off and flying away into the hills, its downwash leaving a long-haired young man struggling to control his jacket and clinging to a microphone. His name flashed up on the screen: Simon Partington.

'Concern is growing for Brian Murphy, a sixty-three-year-old man missing in the hills of Morvan.'

Jen and Rory sat listening in silence as the young man outlined the circumstances of Brian's disappearance. About half of what he said was inaccurate, but he made one thing very clear:

Brian's disappearance had been overlooked by a group of drunken climbers in a bothy. It was their fault and if he was found dead the blame would be all theirs.

Rory picked up the TV remote and the TV screen went black. 'Oh shit. This'll kill Angus.'

Jen squeezed Rory's arm. 'He already blames his sen. Now this.'

Rory got up from the couch and began to pace. 'That's the bloody press. Always looking for someone to blame. They can't get it round their skulls that the mountains are dangerous places and sometimes, well, sometimes things happen.' He stood, shaking, staring out of the window.

Jen walked up behind him and put her arms around his waist. She pressed her face against his back. Jen always loved Rory's back; somehow it was always so firm. 'Maybe they'll—'

Rory cut in. 'No, they won't find him, not now. Not alive anyway. Tonight's his third night. Folk get lost for one night, they maybe break an ankle and they get found on the morning of the second day. But the third night – the third night means ... '

Jen hugged him so hard she could feel his spine, as if she were trying to squeeze the worry out of him. She clung to him for a few seconds, feeling the warmth pass between them like a silent conversation. 'Come on. I reckon we need some scran.'

There is no law in the mountains. They obey no rules and heed no prayers. To those who love them they are magnificent and give life. They give meaning, joy and inspiration, bringing relief from the everyday, the mundane, lifting from us the burden of tedium and giving us glimpses of a world beyond. But those who know them well, perhaps love them best, know that there is a price to pay. It is a blood price; for all this, some will die.

***

At first Angus had expected the phone call at any moment.

'Police here, Mr Sutherland, we've found … '

But the police didn't call. Then the hours that passed became days. Time moved slowly at first but then it gathered pace and seemed to gallop away. Sunday morning arrived – one week since Brian had vanished. Angus was moving some porridge around in his bowl and Laura was in the lounge emailing their son and planning a visit to see him when the doorbell rang.

Laura ushered the police inspector into the kitchen. She was small and blond with sharp features and reminded Angus of a ferret. She was just the wrong side of forty and introduced herself as Josie Redding.

Inspector Redding sat down at the kitchen table with a crisp businesslike manner. 'This is Constable MacLeod.'

The constable was in his mid-twenties, heavy, and with a ponderous air about him. His uniform struggled to contain him. Laura had made coffee almost before the police officers had sat down and she placed a plate of biscuits on the table. Constable MacLeod eyed them with considerable interest.

'Mr Sutherland,' Inspector Redding spoke slowly and deliberately as though she wanted to make sure everyone understood her, 'I've been put in charge of investigating Mr Murphy's disappearance. I've discussed what's happened with mountain rescue.'

Angus knew what was coming. 'Wi' Chris?'

Inspector Redding nodded. 'Yes. We want to thank you for your help. All your people.'

Angus finished the sentence for her. 'But you've decided to call off the search.'

The inspector nodded, Constable MacLeod took a biscuit and Laura stood drying her hands on a dishcloth. Angus nodded slowly. *It's been a week. They cannae search forever.*

Nevertheless the finality of it left an empty space in his stomach. 'I see.' The words sounded like a full stop.

Laura sat down, the tea towel still in her hands. 'We had a reporter round.'

'I expect you would,' the inspector said.

Laura's eyes darted angrily. 'The things he said on the news – that shouldn't be allowed.'

Constable MacLeod took another biscuit and the inspector shot him a disapproving glance.

Angus scratched at his beard, and the inspector placed her hand over his with a sudden and unexpected show of compassion. 'It was no one's fault, Mr Sutherland. These things happen.'

Angus felt his eyes filling with tears and struggled to control himself. The image of Brian's body, cold on the mountainside, flashed before his eyes.

Laura still looked angry. 'You're sure he didn't just go somewhere? He might have. He was *odd*.'

Angus looked up in surprise. 'Laura, you cannae say that.'

Laura drew her mouth into a thin line, like a stubborn child. 'Well, he was. Barely spoke to anyone. You couldn't get a word out of him.'

Constable MacLeod reached out to take yet another biscuit but a sharp scowl from the inspector made him withdraw his hand.

Inspector Redding turned on an official tone. 'None of Mr Murphy's bank cards have been used and no one saw him leaving on the ferry. These facts suggest that he's still on the island.'

As the weeks passed, the press went quiet and life slowly returned to normal – and it carried on without Brian. For most people it was as though he had never existed, but not for Angus. He stopped going to the mountaineering club. He couldn't stand to be in the Thistle Inn with an empty chair where Brian had been, and he couldn't face jovial discussions about someone's next Munro. It all seemed too trivial. He made excuses when Rory called to suggest trips to the hills, and spent his weekends in the garden. The only distraction he could find now was his work, so he immersed himself in it, working late and even going in at weekends. Sometimes he made work up when none existed.

***

It was five-thirty and the council offices were mostly deserted but Angus felt no urge to go home. He was preparing for a big council meeting a week on Thursday when he noticed the Muir estate's application for a licence to release a pair of lynx back into the wild. With everything that had happened with Brian, Angus had forgotten about the application. He opened the large folder and took out the papers. The application was lengthy and well researched, and the estate gave assurances that the lynx would be tagged and monitored. Furthermore, since Morvan was an island and completely cut off from the mainland – except for the ferry link – the reintroduced animals would be isolated in a comparatively small area.

A number of environmental groups supported the application: Friends of the Earth; Scotland: The Big Picture; and the RSPB. He read the list of objections, one by a group of local shepherds fearing loss of livestock, and a second – predictably – from Lord Purdey's Campaign Against the Persecution of Landowners. Purdey's complaints ran to several pages and predicted that Highland agriculture would collapse under the attacks of wild lynx. Angus chuckled at that. According to blood-sports enthusiasts almost anything could bring about the downfall of their way of life. Angus ran his finger down the list of councillors who were to vote on the application. Over the years he'd learned to predict with unerring accuracy which way each of the elected members would vote.

'For, against, for, for, against,' he muttered under his breath.

The vote was going to be close. Some might be swung one way or the other by the debate in the chamber; others would have made up their minds long before they took their seats. There was, however, one key vote: the chairman of the committee, Gerald McCormack. Angus could see him in his mind's eye – a tall, imposing figure who enjoyed the sound of his own voice and could swing the committee either way if he had a mind. And Angus knew that Gerald McCormack was Purdey's solicitor.

'Against,' Angus said as his finger passed the chairman's name. 'McCormack will vote against. So the application will fail.'

Angus remembered his conversation with Rory and Sinead in the bothy. He remembered her passion for the return of the lynx, how Rory had argued for it too. Then he visualised Glen bothy surrounded by woodland and lynx moving between the trees like ghosts. He hadn't been back to the bothy since they'd lost Brian. Something in him couldn't face the place, and the urge to go on the hills had faded since the accident.

Then Angus thought about Rory's attempts to teach him about wildlife, about the delight he'd felt watching bats at dusk and stamping down bracken to help rare and special creatures thrive. Nature *did* deserve a chance – Rory and Sinead had helped to open his eyes, and now maybe he could do something too. If Angus did nothing else with his time in the council, he'd make sure the resolution over the lynx would pass. If there was a way, he'd find it.

It was after six-thirty when he began his slow walk home. The town seemed quieter than usual as he walked over the bridge across the river and up the high street. As he passed the Gellions pub the doorman lounging outside noticed him and waved. Music spilled out and he heard the chatter of voices inside. He found himself crossing the street and walking through the door. He had passed this pub a thousand times on his way home and never thought to walk in, yet here he was, leaning against the bar surrounded by drinkers as the jukebox blared out Tammy Wynette standing by her man. The bar was long and narrow, its wood-lined walls hung with the usual array of fake old pictures of towering mountains and fierce tartan-draped clansmen.

'What can I get you?' The barman's voice cut through the noise. He was in his early thirties, with a dark neatly trimmed beard and tattooed letters running around his neck.

Angus stared at him. He'd not thought about what he wanted to drink. It was early, and he never drank anything alcoholic before nine in the evening.

'An orange juice, please.'

The barman turned and headed for the chiller cabinet, but Angus called after him. 'Actually no. Make it a large Bell's.' Angus scanned along the rack of whiskies and his eyes fell on a single malt he liked. 'In fact, make it a Bunnahabhain, a large one, yer ken.'

He would treat himself – why not? He smelt the glass first, savouring the dark peaty aroma. It reminded him of the hills and smoky bothy nights and the whisky warmed him as he took it down, filling an empty space inside him. His phone vibrated; it was Laura. He switched it off and ordered another whisky.

It was ten o'clock before he finally put his key in the lock – it took a couple of attempts – and shoved open the kitchen door.

Laura was waiting for him. Her expression told him that she was more worried than angry. 'Where have you been?'

An intense, deep-brown smell filled the kitchen. He guessed it might once have been beef casserole. Laura had probably kept it warm for hours.

He wasn't drunk, just a little merry perhaps. 'I was working. On that application to release lynx.'

Laura looked at him with growing amazement. 'Have you been ... drinking?'

'I called in to the Gellions for a couple.'

Laura took the casserole dish out of the oven and spooned some thick, meaty gloop on to his plate. He expected her to be angry, but she seemed just to be puzzled. The plate sat steaming in front of him. He forked some of the brown sludge into his mouth. Surprisingly, it tasted much better than the usual seared flesh Laura served up.

'It's good,' Angus announced.

Laura was staring at him from across the table. 'So, you were telling me how you have been working on the lynx application.' There was an odd tone in her voice.

'Aye, there's lots tae do,' he lied. 'Imagine bringing lynx back to the forests of Morvan after 2,000 years. Incredible to see those creatures back. Oot in the wild again.'

Laura's face wrinkled. 'You were never interested in lynx before. Anyway, I don't think having those things running around would be a good idea. I'm not sure I'd feel safe.'

Angus laughed, and words tumbled out of him. 'Yer'd never see them. "The Keeper of Secrets", that's what they call lynx. They'd just vanish into the forest.' He felt as if he could eat another plate of stew. 'Is there any more?'

Laura didn't speak, but he could see the surprise on her face – he never asked for second helpings. Silently she spooned more of the remains of the casserole on to his plate and put it down for him.

'Be sensible, Angus,' she said quietly. 'Surely they'll carry off sheep.'

Angus waved his fork in the air to emphasise his words. 'I read the application – they reckon we might lose four or five a year. Lynx are reluctant to leave the forest so sheep are fairly safe.'

Laura was looking at him with growing amazement. 'You'll be wanting to bring the wolf back next.'

Angus rolled his eyes, and then thought that the wolf might not be a bad idea as well. 'Aye, why not? Let's have wolves too. Rory says they did it in Yellowstone and everything was fine.'

Laura looked increasingly irritated but the whisky made Angus feel comfortable and once he started talking he didn't think he could stop.

'There have been no recorded wolf attacks in the USA for fifty years. Wolves dinnae attack people.' He grinned. 'And, for that matter, recorded accounts of them dressing up as grandmothers are actually pretty rare. I have heard that bears will eat porridge, although I believe they are in fact not that fussed about the temperature.' Angus giggled.

Laura looked at him as if he'd taken leave of his senses.

Angus was laughing now, almost falling out of his chair. 'But it is actually true that the bears are uptight when it comes to little girls sleeping in their beds, yer ken. That's a no-no for bears, that is.'

'Oh, for God's sake.'

Laura picked up the empty plate and thrust it into the dishwasher where it clattered into the rack. She was walking away into the lounge when Angus called after her.

'This is only the council approving the application, mind. They send a recommendation on. After this it has to be approved by the Scottish Government.'

Angus heard the television come on in the lounge. He sat quietly for a few minutes, hands palm down on the table. When he looked down he noticed that they were trembling.

***

It was early afternoon when Jen woke. The hospital smell from last night's shift still hung in her clothes. Rory had gone to work. She'd not heard him go; he had slipped away quietly as he always did.

After eating some muesli, she ran the hoover over the living-room floor. She was about to put the hoover away when she found herself standing beside the old pine bookcase Rory had rescued from a skip. She pulled out an old battered textbook, *Clinical Nursing Procedures*, and it fell open where a brown envelope was sitting between the pages. Jen picked it up and stood staring at it for a moment.

The letter had arrived a few days ago. She remembered how shocked she'd been when she'd opened it and found the solicitor's grandly headed notepaper telling her of her Aunt Geraldine's death. She'd been even more surprised when she found a cheque for 80,000 pounds neatly stapled to one corner. She hadn't told Rory yet. It was enough money for him to have his island dream, for them to set up in a remote self-sufficient haven – the trouble was, she wasn't too sure it was her dream as well. It would mean leaving the hospital, leaving her whole career behind. She didn't know if she could take that step. She put the envelope back into the book and slipped it back into the bookcase.

She'd tell him soon, perhaps tonight or tomorrow. He'd be delighted, she knew that for sure, and that was the problem. She had a dim memory of Aunt Geraldine, maybe from a family wedding. She had stunk of gin and had bad teeth. That was about all Jen could remember. How incredibly thoughtless of the old lady to die and turn Jen's life upside down. Rory's dream had been something he'd clung to, but it had remained a dream – 80,000 pounds made it frighteningly real.

The door buzzer shook her back to reality. She picked up the phone receiver to talk to whoever it was three floors below.

'Jen, it's Laura. Can I come in?'

***

Laura sat in the small lounge cradling a coffee. Jen couldn't remember the last time she had said more than a few words to Angus's wife, and she sensed that Laura had come for something important.

Laura glanced out of the window that looked over the Ness. 'You've a lovely view here.'

Jen smiled, knowing that Laura's comment had only been polite small talk. 'Aye, well, it suits us.'

Jen waited and let the silence ask the questions for her. Laura sipped her coffee, looking about as comfortable on the settee as if she were sitting on a nest of ants. 'Angus, he's … '

Laura's words stalled. Jen sat and waited.

She began again. 'He's come home late a few nights. Called in to the Gellions on his way home.' Laura paused to let her words sink in. 'Angus, in the Gellions! It's not his sort of place at all.'

Jen nodded. 'He's not been to the mountaineering club neither, tha knows.'

Laura nodded. 'Since Brian … ' Laura let the word hang in the air. 'He's not been himself. Goes to work but that's all.'

'Aye, he's not his sen.'

Laura sat with her head bowed, avoiding Jen's gaze. 'I never much liked him going out on the hills. Off to all these bothies, coming back stinking. At his age.'

Jen smiled.

'I used to go with him. Years ago. But that was before ... '

Jen felt her stomach turn as she noticed that Laura's body was trembling, as if there were something she couldn't speak about.

Laura carried on, leaving whatever it was unsaid. 'But without the mountains, it's like he's lost. He doesn't know what to do with himself.'

Jen nodded slowly. 'Aye, Rory said something like that.'

Laura had reached the point of her visit. She spoke more firmly now. 'I want Rory to get him out. Climbing or something. He's got to get out.'

'Rory's had a go. Angus allus makes an excuse, like.'

Laura was suddenly determined. 'Will you ask Rory to call him again? Arrange a trip?'

Jen nodded. 'Aye, I'm sure he will, but Angus maybe'll nah go.'

'You tell him to ask. I'll make sure he goes.'

After Laura left, a small pool of tension remained in the room that only slowly dispersed.

***

Rory couldn't help but notice that Angus was quiet on the walk up the glen. Rory had joked on the way in, chatted, even whistled, but Angus remained stoic and glum. As they rounded the corner of the glen the ridge rose imposingly before them like a giant shark's fin.

Rory dropped his pack and wiped the sweat away. 'There

you go, Angus. What do you think?'

Angus looked up. 'Aye, spectacular.'

It was high summer now and the hills were alive with calling birds and the constant hum of insects. Rory pulled out his binoculars and searched the meadow grass. 'I've got him.'

Angus put down his water bottle. 'What is it?'

Rory peered through the glasses. 'Male stonechat. Colourful little soul.' Rory passed the binoculars to his friend. 'See, on that little bush by the side of the path.'

Angus spotted the little black-headed bird with its rust-coloured chest sitting high on the bush. 'Aye, I ken him now.'

Angus looked interested. It was the first time Rory had seen him show any enthusiasm since Brian had vanished.

'Listen to his call. Sounds like stones knocking together. That's where the name comes from. Stones chattering.'

'Ah, that's braw. I've heard it many a time but nivver ken what bird it was.'

Angus watched the little bird for a few minutes until eventually it flew off. At last Rory could see him taking an interest in the things around him. A little of his old spark seemed to have returned.

Three hours later, Angus and Rory were on the last pitch of the Great Ridge of Ben Bhuidhe, high on the sharp ridge with the green glen laid out before them. The smell of summer carried up to them on the wind. Angus had climbed well on the route, Rory thought; even though the older man hadn't led it, he had quietly followed up and over the steep rock pitches. But despite this, Rory had the feeling that something was still distracting Angus. Rory completed the final pitch of the climb, a series of mantelshelf moves leading to increasingly easy ground.

Rory secured the rope to a huge boulder at the very top of the ridge, wedged himself into a small gap, relaxed and began taking in the rope. He felt it go tight.

Angus called up: 'That's me, James.'

Rory frowned, unnerved. He held the rope tight. He couldn't see Angus below him, only feel the rope moving slowly, sometimes jerking as Angus made a move. The ridge was not a hard climb for Rory but Angus struggled on the last fifty feet and Rory kept the rope tight to give him confidence. Rory gazed out over the landscape, watching huge grey clouds swell from the west like galleons under sail. A raven's harsh call echoed back off the rocky cliffs. Rory watched a pair of the black birds circling below, weaving in and out of the thermals, their bodies so dark they seemed impenetrable to sunlight. That sound – that echoing caw that seemed as hard as the rock itself – was the sound of high places, the sound of the ridges and high corries, a lonely wild sound. *This place belongs to them*, Rory thought. *We are only visitors.*

A hand appeared over the lip of rock at Rory's feet. Then it was joined by another hand, with a red-faced Angus swiftly following.

'I'm getting too auld for this, James,' Angus said to Rory, slowly getting back his breath.

On the safety of the summit, Rory began packing away their climbing equipment. The rock's roughness lingered in the memory of his fingertips, and the balls of his feet revelled in the motion of walking after being restricted to the small holds of the rock face for so long. The island stretched away below them.

Smaller islands along the coast showed green in the glittering blue sea. Sometimes Rory forgot why he had turned to climbing, but on days like this, high on the mountain with his friend by his side, it seemed absurd to do anything else.

Rory waited until Angus was relaxed and smoking his pipe. He had to choose his moment carefully, but he had to ask. 'Angus, who's James?'

Angus looked up, confused. 'James? What do you mean?'

Rory dumped his rucksack and sat down beside his friend. 'You called me James.'

'Nae, I dinna. Did I?'

'Yes, Angus, you did. Twice.'

Rory watched as Angus puffed out a cloud of pipe smoke and pretended to be looking at the view.

'Jis' a slip o' the tongue, yer ken.'

Rory asked again – slowly this time, so Angus would realise he knew that there was something more to this. 'Who is James, Angus?'

Angus looked away quickly. 'Jist someone I climbed with years ago.'

'Back in the bothy, the day Brian got lost, you called the rescue team leader James before you ... before you kind of passed out.'

Angus shrugged and stared at the rock below him.

Rory followed his hunch. 'What happened, Angus?'

Angus was shaking, gripping his pipe as hard as he could, but at least this time he didn't try and make an excuse.

Then Angus spoke, so quietly that Rory could barely hear him. 'It's a long time ago. More than thirty years, yer ken.

We climbed all over, him and me. We were on Ben Nevis and I wanted to climb Green Gully. When we got there it was thin. Unclimbable, I should ha' ken. I was in front an' I got this bit of ice I couldnae climb. He offered tae try it oot. He was better than me, yer ken, but I was stubborn. I almost did it, but I fell.'

Angus fell silent. Rory put his hand on his shoulder. 'It's all right, don't tell me if—'

Angus cut in. 'Nae, if I dinnae tell ye I never will. I pulled him off. His belay failed. I broke ma hand but he … he … '

Angus sobbed quietly, and Rory squeezed his shoulder, unable to find any words.

Angus looked up and his eyes filled with tears. 'He looked so perfect, yer ken. Just some blood in his nose. That wis all.'

Rory stroked Angus's shoulder. 'It's all right, man.'

Angus shook his head slowly. 'Nae it's not all richt – it'll nivver be richt. James was Laura's brother.'

Rory watched the pain welling inside Angus. Sometimes all you can do is wait beside another human being in silence, with your hand on their shoulder, while the storm rages. Angus was quiet on the way down, but he seemed more at ease with himself now that he had told Rory about the memory that tortured him from thirty years ago.

Rory set off in the lead. He was carrying the bulk of the climbing equipment, the rope coiled on the top of his rucksack, but he had a grace on the hill, and moved with deer-like ease down the track beside the gorge as they headed back to Glen bothy. The wind picked up as they descended and soon the trees in the valley floor were sighing and shifting, a sea of green oak.

They often saw the hen harrier now. In August, it would

have young to feed, and so was constantly hunting for as long as the daylight hours would allow. Half an hour later Rory and Angus saw the familiar flicker of the white wing with its dark tip weaving across the landscape only a few feet above the ground. The male was hunting again, seeking out voles and small birds hiding in the tussock grass. Angus and Rory lay in a hollow in the hillside and passed the binoculars between them, taking it in turns to watch the harrier.

Suddenly the elegant bird dropped into the grass, only the tips of its wings betraying where it lay.

'It's made a kill.' Rory was peering intently through the glasses.

The bird rose with something in its beak. Soaring high, it turned to head down the glen with its catch, but a loud crack split the air and the bird froze for a second. Then it plummeted to the ground and vanished into the grass. Rory felt a moment of disbelief, followed by something approaching panic. *What the hell?*

He was up on his feet and running towards where the bird had fallen. Angus followed. They ran hard and crested the ridge, searching in the grass, but saw no trace of the downed bird. On the far side of the ridge, the two Purdey ghillies were heading away, one carrying a shotgun, down towards their Land Rover on the track a hundred yards below.

'Bastards!' Rory yelled. His voice carried across the corrie.

The keepers heard him and stopped, looking back up the hillside to the two walkers.

'Get a photo, quick,' Angus cried.

Rory fumbled for his phone and raised it to his face. The ghillies saw the phone and turned, hurried to their Land Rover and drove away.

'Did you get them?'

'Aye.'

It took them half an hour to find the bird. They had watched it cut through the air like a wild wind-filled spirit. Now it lay lifeless and twisted in the grass, the vole gripped in the harrier's orange beak. They photographed the bird and Rory used his GPS to pinpoint their position. Finally they gathered the lifeless creature and placed it in a rucksack.

When they headed back for the bothy, the glen was empty and still.

CHAPTER 9

Rory looked around Sinead's office. It was in a small courtyard hidden away behind the Muir family's grand estate house. The house didn't impersonate a medieval castle like the Purdey mansion, but it was dominated by a large tower at its centre. The tower, topped by a spire, gave the impression of an ornate English church dumped into the Highland countryside for some reason. The Muir family had bought the house and the land around it some decades ago – which meant, by the usual standards of Highland estate owners, that they had just moved in.

Sinead's office was small but comfortable, with a couple of faded armchairs and an old desk that supported her work computer. On their way here, Angus had told Rory about the estate offices he'd been in before. Normally they were hung with deer heads and the sporting trophies of shoots. Photos of shooting parties would grin from the walls, and gun cabinets would be prominently displayed. They were places where the last vestige of the Victorian age hid away, clinging to power in the vast open spaces of the Scottish Highlands. Only in the Highlands do a handful of men hold sway over huge areas of land and have near-total power over what happens in the rugged mountains and wide glens. Whenever Rory looked at these images, and the heads of stags hanging with dead eyes from the walls, he would shiver, wondering why these men were obsessed with turning the hills he loved into killing fields.

Sinead's office was different. Her walls were decorated with images of life, celebrating the creatures which roamed or had once roamed these mountains. There were pictures of beavers, eagles and hen harriers, and a bookcase crammed with books about flowers, trees, fish and birds. Rory's eyes were drawn to the photograph above Sinead's desk. It was a framed portrait of a lynx. The big cat was hunting through a pine forest, its ears high and alert, adorned by tufts of hair that sensed the wind direction, its body lean and taut. The photograph had captured the cat in an intense moment – perhaps catching the scent of a roe deer, or hearing the soft footfall of a wolf. Its fur was mottled brown and black, colours evolved over thousands of years to make it invisible in the dappled light of the forest. Most of all he noticed its eyes, alive with a golden light that shone out of the forest. He felt as if the animal were looking directly at him, calling to him from a long way off; far away not in space but in time.

There was a note on the desk that read 'Back in five minutes'. They didn't have to wait that long, though, as moments later the door opened and Sinead entered the room, her arms laden with files.

'Och, Jasus – you fair startled me.'

Angus cleared his throat. 'We thought yer ought to see something, ken.'

Something in Angus's tone made Sinead stop in her tracks. 'Aye?'

Rory showed her the bag with the dead bird in it.

Sinead walked slowly to her desk, but she didn't sit down. She kept her eyes fixed on the bag. 'What have you there?'

Rory put the bag down carefully on the desk. 'It's a harrier.'

Sinead had a look of horror as she opened the bag and peered inside. 'Would you look at that now. Where?'

Rory hesitated; it was Angus who told her. 'We saw it shot, up on the Black Moor.'

'Oh mother of God.'

Sinead reached into the bag and lifted the bird gently out. It lay in her hands, a twisted bloody mass of grey and black feathers, its head lolling to one side and beak hanging open.

Rory watched her lay the bird down on the desk. He noticed how she touched it with immense care and reverence, as though handling some fragile ancient artefact. Sinead slowly spread out the wings, smoothing the grey feathers out with sensitivity as if, by her care, she might restore the creature to life. There was something in the way her fingers pushed the feathers back into alignment and straightened out the twisted neck that reminded Rory of how the skilled hands of a watchmaker or artist might bring something to life. It took her ten minutes, but she did not rush. At the end the bird lay before them on Sinead's desk looking almost perfect – apart from the bloody holes, and the fact that the wings, which had captured the wind in life, were still and dead.

Sinead stepped back and looked down at the dead hen harrier. Anger boiled behind her calm expression. 'An' ye feckin' *saw* this, did ye?'

Rory nodded. 'Yeah. We were coming down the ridge, saw the bird, heard the shot.'

Angus lowered himself into a chair. 'Aye, we saw the bird fall deid. Then we saw the keepers getting into yon Land Rover.'

Sinead picked up her phone and dialled. 'Tony, where are ye? There's something here ye'd better see. Aye, I'm in the office.' She put down the phone. 'These fellas, did ye know 'em?'

Rory could see the anger in Sinead – it was raging and she was struggling to contain it.

'I've seen them before. I don't know their names.'

Angus thought for a moment. 'One was a grey auld man wi' a deerstalker on and the other were a big ginger loon.'

'Aye, the oul man's Donald and the ginger eejit is Hamish. Purdey's ghillies.' Sinead stroked the bird slowly.

The door opened and a short scruffy man with a long grey ponytail came into the office. He was dressed in shorts and wellington boots, and was about to wipe his feet when he saw the harrier and his eyes widened.

'Oh my God. What happened?'

Sinead hissed, as if she could barely let the words escape. 'Purdey's keepers shot it. These fellas saw it.'

Tony held out his hand to shake Angus's hand but hesitated, perhaps realising that it was caked in mud. 'Sorry, I was working.'

Sinead introduced them. 'Tony, these fellas stay in the bothy a lot. Angus and Rory.'

Angus smiled. 'Fit like. Yer the gardener here, then?'

Rory laughed and Tony looked a little embarrassed. 'Actually I do work in the garden, but I own this place.'

'Angus, Tony Muir,' Rory said quietly.

Tony beamed. 'I'm the rich bastard who owns this place but try not to hold it against me, man.' He turned to Rory. 'So you stay in Glen bothy, do you?'

'Yes, quite often. I suppose we sleep at the bottom of your

garden.'

This seemed to amuse Tony. 'Well, it is a pretty big garden. This is a bad business, though.' He looked down at the dead harrier.

Suddenly Rory felt a sadness welling up inside him. 'It was awful when we found it. All twisted and bloody. Such a beautiful bird. Bloody gamekeepers.'

'So you'll report it, then?' Sinead asked the hillwalkers.

Rory didn't wait to ask Angus. He was full of rage. 'Yes, of course.'

Sinead picked up the phone. 'I'll phone the local RSPB fella for you.'

She held the phone for a moment or two. 'Answerphone. I'll go straight to the police, so I will.'

While Sinead was dialling, Tony gestured to Angus and Rory to follow him. 'Come and take a gander at this.'

Sinead looked up from the phone. 'They'll not be wanting to see ... '

Tony grinned back at her. 'Of course they will.' Out in the yard he turned and whispered, 'She doesn't understand these things.'

Angus talked as they walked across the yard. 'I was reading your application to release lynx yesterday.'

Muir looked surprised. 'Really?'

'Angus works for the council. Deals with all the minutes of meetings and stuff,' Rory explained.

'Ah. What are our chances of getting it through, do you think?'

'Well, there are a few votes you can count on, but unless

you can get Gerald McCormack on board yer'll struggle. I hear he's a big pal of Purdey's so we know who he'll side with.'

Rory wanted to see one environmental measure get through. 'Highland Council never stand up for nature. Look what they did to Glen Etive.'

Angus shook his head slowly. 'They approved *seven* hydro schemes there. It's a National Scenic Area but that made nae difference. And there were protests, yer ken.'

By now they had arrived at a huge green garage door and Rory watched as Muir fiddled with the combination lock.

'Money,' Tony said, 'that's all the council cares about. They just think if it'll make a few bob then it's OK – sod the environment.'

Angus coughed, as though he were making an announce-ment. 'We thought we might write a letter of support. The mountaineering club, I mean.'

Muir had the lock off at last and beamed his appreciation. 'That would be very kind. I think we'll need all the help we can get.' He sighed. 'There's so much prejudice about the lynx. Problem is, people don't know anything about them – they're *scared* of 'em, you know. Can't bear the thought of something big out there with teeth and claws creeping around in the night.' Muir raised his hands to his face like the claws of a great cat and roared.

Rory found himself wishing that there were more landowners like Muir, but he knew that Muir was one man; the rest of the 400-or-so folk who owned most of the Highlands considered the carnage they caused not only normal but laudable.

'That's the trouble with the nanny state, you see,' Muir continued. 'People want the world all sanitised and safe and it shouldn't be. The world is far too regulated.'

Angus smiled. 'Maybe we could all do with a bit more wildness, yer ken.'

Muir grabbed Angus gently by the arm. 'Quite right.' Then he drew him closer. 'Take a look at this.'

With a theatrical flourish Muir slid back the big garage door, which rattled aside on castors. A strong smell of oil and steel drifted out of the darkness of the garage.

Rory peered into the gloom.

At the back of the garage he saw a huge green machine. It sat astride enormous caterpillar tracks, the massive barrel of its gun pointing skyward.

'My God, it's a tank!'

Muir grinned. 'A self-propelled gun, in fact, from World War II. Had it restored.'

Angus walked slowly into the garage, peering at the enormous steel tracks and examining the barrel. 'I dinnae ken you were allowed to have things like this.'

'Oh, you can have a tank if you like. Not so keen on you having ammunition for it though, sadly.' Muir sighed.

Rory viewed the hulking steel beast from a safe distance, wondering why anyone would want to own such a thing. 'What do you do with it?'

'I drive it around the estate now and again. I'm not supposed to be on the road with it. Lots of regulations. I used to have an eighty-eight-millimetre howitzer, which sat up there just by the main entrance.' Muir pointed up to the turreted house.

Angus was stroking the tank, as if it might wake at any moment. 'But you've nae got it now?'

'No, chap from the council came round. Apparently there's

some bloody by-law against having light artillery in your front garden. It wasn't as if I was going to fire the bloody thing. Although between you and me I did think November the fifth might be an opportunity. I thought I might put a shell over the river. Kiddies would have loved it.'

Muir laughed loudly and it took Rory a moment to realise he was joking. 'Yes, I'm sure that would have gone with a bang.'

Sinead appeared behind them. 'Don't mind him. He's not quite as mad as he pretends to be.' She glanced at her watch. 'It's four o'clock, Tony. Aren't you meeting those fellas about the roof?'

Muir turned with a start. 'Hell yes. I'd completely forgotten. Have to go.'

'We'll write that letter in for you,' Angus called after him.

Muir waved over his shoulder. 'Thank you.'

Rory watched the estate owner jog away in his wellington boots.

Sinead turned to Rory and Angus. 'He's no a bad fella. The polis want to speak to ye, get a statement, so they do.'

Rory was pleased that he could do something to get back at the keepers who had shot the hen harrier. 'Great. We've got photos too.'

'Before you go, there's one thing.' She looked solemn. 'Purdey's solicitor came round just now, that McCormack fella. Unless we can find those deeds in the next few weeks he's going to start proceedings to get the land back. And we've looked everywhere. Nothing's turned up.'

It seemed so unjust to Rory that anyone could close a bothy on a whim when they were such important things to so many people. 'No leads?'

Sinead shook her head. 'Tony can't remember what happened to the deeds. It was thirty years ago. An aul' shepherd lived in the bothy, but that must be decades ago and I don't know where he is now.'

***

Angus swung the Land Rover up the drive and on to the road out of the estate. 'Imagine having your own tank, yer ken.'

Rory laughed. 'If you've got as much money as he has, I suppose you can have what you want.'

They drove in silence for a while, and then an idea came to Rory. 'A shepherd lived in Glen bothy, Sinead said. I tell you who might know about him.'

'Who's that?'

'Christine who runs the shop. There's nothing she doesn't know.'

***

Morvan village shop sold everything – or at least it sold everything anyone could possibly need. If it wasn't in the shop, then you didn't need it. But variety was not something the shop could supply. There were only two types of sausages and one kind of meat pie. Shelf space was limited to one or two of everything. Paint cans jostled with tins of soup. Bread sat next to cakes and newspapers. Whisky sat next to orange juice. Despite the shop's limited range, it would be perfectly possible to live your entire life just from the contents of those shelves.

Christine was a small, plump woman in her early sixties.

She wore several pullovers and an apron tied loosely about her waist. Rory and Angus walked up to the counter; they could see Christine sorting cans in the small office beyond.

Christine had heard them come in, but she let them stand there for a few minutes before she put down the tins and emerged from the storeroom. Christine did not hurry. There was no need. Her customers had nowhere else to go if they needed bread or milk, so she took her time, helping them to understand that there was no benefit in life to be had from rushing.

At last she sailed up to the counter. 'Now then, what can I be getting you?'

Rory began the quest. Not quite knowing where to start, he stammered, 'We stay in Glen bothy quite a bit, you see.'

Christine nodded. She knew this already. Second sight or supernatural powers would have been wasted on Christine; she did not need them. She already knew all there was to know about the village and its inhabitants. When people came into the shop they passed the time of day, and by doing so gave away odd facts about themselves. Christine filed these facts in her computer-like memory, along with gossip she had heard the previous day or a rumour the postman had told her. She combined this information with her encyclopaedic knowledge of the shopping habits of the 200-or-so folk who were regular visitors to the shop. She knew that Mrs MacLean was very fond of Milk Tray, for instance. Mr MacLean never bought Milk Tray, but when he went overnight to Inverness every other Saturday, the village postman – whose name was William and had an eye for the ladies – bought a box without fail. Now why might he do that, do you think?

'Oh yes.' Christine smiled sympathetically. 'And you'll be

the vegan.'

Rory was surprised she even knew him. He had only called into the shop half a dozen times. 'Why yes, I am.'

'Never buys meat.' She shook her head sadly and turned to Angus. 'That'll be why he's so thin. A bacon sandwich would do him the world of good.' She smiled at Angus. 'Will you be taking a drop of your usual?' she asked and simultaneously placed a half bottle of his favourite whisky on the counter.

Angus bought the whisky and as he paid for it asked the woman if she knew the old shepherd who had lived in the bothy.

'Oh of course, he was some worthy, that man,' she replied, placing the bottle in a plastic bag and handing it to Angus.

'He's dead now, I suppose?'

'Oh, dear no,' she exclaimed. 'He's some bodach but he's certainly not dead. He's at the old folks' home in Inverness.'

Rory wondered why the old shepherd would move all the way to Inverness to be looked after in his old age. 'Inverness?'

Christine was on her way back to her tins. 'His son lives there. Nice young man but a bit too fond of super lager. Ask them for Mr Campbell.'

'Thank you.' Angus turned to leave.

'A few rashers of bacon just to keep out the cold, young man?' Christine called as Rory headed out of the door.

Before they reached the door it burst open and the tweed-suited figure of Tabatha Purdey entered the shop with all the force of a Highland clan charge.

She collided with Angus and almost knocked him off his feet. 'Terribly sorry, old thing. Not looking where I'm going.'

'I'm fine. Nae bother,' Angus answered quickly.

Christine called over from the counter, 'Gin, Tabatha?'

Tabatha began to head for the counter and then turned back to Angus. 'Have we met?' She turned to Rory. 'You both look terribly familiar.'

Rory knew he'd heard that voice before but couldn't place the solid tweed-clad woman in front of him. 'No, I don't think so.'

Tabatha peered at him. 'Are you quite certain?'

When she spoke those words Rory's mind brought back a memory. He had been standing in the darkness beside the river, naked and reaching for his dry clothes, when a blinding light had flooded him. Then he'd heard a woman's voice: 'Well, I suppose it's better than fairies.' It was the same voice talking to him right now.

Rory glanced at Angus and could see from the horrified look on his face that he too remembered that voice. Rory found himself struggling to suppress a giggle. He raced for the door. 'I don't think we've met,' he stammered.

They bustled out of the shop, desperate to make it to the Land Rover before they were recognised.

Rory burst into laughter as they ran across the car park. 'She couldn't recognise us with our clothes on.'

They fell silent on the drive out of the village and over the high mountain pass that led to the ferry back to the mainland. The trauma of seeing the harrier shot had driven Angus's words on the climb out of Rory's mind, but now he played their conversation over again. He understood how haunted Angus was by the death of his brother-in-law all those years ago – a hard thing for anyone to bear – and Angus's anguish at Brian's disappearance was easy to understand. Such an awful thing, to be transported back to that icy face and to relive the death of his friend – every climber's nightmare. Rory wasn't sure that he would climb again if he had that to face.

He glanced over at Angus, who was focused on the driving, and wondered if talking about the accident had made things better or worse for the older man.

Rory thought he should say something now that they were alone in the Land Rover – perhaps give Angus a chance to talk. 'Angus, I'm sorry about James. It can't be easy.'

Angus didn't reply at first, his eyes fixed on the single-track road. 'Aye. Well it were a lang time ago, yer ken. Jis' now and again like. Laura … ' The words stuck in Angus's throat. Rory waited. 'I dinnae ken as she'll ever be the same.'

Rory tried to pick the right words but everything sounded wrong in his head. 'I know it must be hard but that's climbing. There's always accidents. It's part of the game.'

Angus gripped the steering wheel so hard his knuckles had turned white. 'Aye, I ken climbing. It's part o' the game. But sometimes you lose big, like.'

Rory started to answer but Angus put up his hand. 'I ken yer trying to help. But let's jis' leave it the noo.'

The front wheels of the Land Rover hit the loading ramp of the *Maid of Morvan* and they crossed into another world.

***

Hamish leaned on the bonnet of the Land Rover and his weight made the vehicle's suspension sag. Donald was in the driving seat, the smoke from one of his evil roll-ups drifting into Hamish's eyes. Purdey, dressed in an oversized oilskin coat to ward off the drizzle, paced the gravel drive and they all eyed the open gate of the estate waiting for the police car.

Hamish watched his employer pacing on the gravel, noticing how his brogues left small indentations at each step. The hillwalkers had popped up out of nowhere when he'd shot the hen harrier. He still couldn't understand how neither he nor Donald had seen them.

Purdey shook the rain from his collar. 'They won't be able to prove a damn thing.'

Hamish wished he was so confident – it had been as close to being caught red-handed as it could get. He glanced at Donald, who looked calm and inscrutable.

Donald tossed his cigarette butt into a puddle where it hissed for a moment before going out. 'Hawd on. Here they is, by the way.'

A white-and-yellow squad car was driving slowly through the gates. Purdey set his jaw ready for battle. 'Now then, let me do the talking. You agree with everything I say and we'll get rid of these plods in no time.'

Donald stepped out of the Land Rover. 'Aye, sir. Youse can honnel it.'

Hamish straightened up and the Land Rover groaned as his weight came off the suspension. 'Aye, right enough.'

***

Hamish tried to look invisible against the wood-panelled wall of Castle Purdey's estate office, but it's difficult to become invisible when you weigh seventeen stone.

On the other side of the room, Inspector Redding was sitting bolt upright in the chair, being peered at by a large number of disembodied stags' heads; some of them even looked curious as to what she was about to say. Hamish placed a tray of tea and biscuits on the desk and Purdey leaned back in his chair, beaming as if he owned the place (which, in fact, he did). Donald and Hamish sat against the wall in a vain attempt to become invisible. The big police constable unbuttoned his uniform and took a digestive biscuit from the plate.

Purdey was all smiles. Hamish couldn't help feeling surprised by how incredibly pleased the laird was to see the two officers of the constabulary.

Purdey was just complimenting the police inspector on how smart she looked when she flipped open her notepad and explained that she had had a complaint about a harrier being shot.

The landowner raised an eyebrow. 'Really, a hen harrier, shot? How dreadful.' With an air of complete astonishment Purdey turned to the keepers. 'You wouldn't know anything about this, would you, boys?'

Hamish and Donald were equally astonished.

The inspector fixed them with an unyielding gaze while she read out the descriptions of the two men seen on the moor. 'One youngish, large and thickset with ginger hair.'

Hamish sucked in his stomach, and, if he had been able, would have sucked his hair in too. The constable took another biscuit and the inspector glared at him.

Inspector Redding read more from her notes. 'An older man with grey hair and a beard wearing a deerstalker.'

Donald glanced down at the battered hat in his hands and

then looked piteously at Purdey. The inspector closed her notebook and fixed her eyes on the keepers.

Purdey stared at the pair incredulously. 'Good Lord, that's you two to a T. Isn't that remarkable?'

The keepers nodded in unison and Hamish wanted the ground to swallow him. 'Ah, wait,' Purdey cried. 'The Black Moor ... didn't I send you up there shooting hares?'

'Hawd on,' Donald said after a moment. 'So we were!'

Hamish nodded. In the past he had found that the more he spoke the more likely he was to get himself into hot water, so he adopted the policy of saying as little as possible. 'Aye, right enough,' was his response.

The inspector had her notepad out again. 'Shooting hares. Is that why you were up there?'

Purdey answered. 'Oh yes, terrible problems with hares. Infested with ticks, you know, and they infect the grouse with disease. Have to keep on top of the population.'

The inspector wrote that down in her book. 'Witnesses say they heard a shot, the bird went down, and you were seen leaving the scene.'

Purdey grinned. 'Ah, so these hillwalkers—'

Inspector Redding pounced. 'I didn't say they were hill-walkers.'

Hamish felt beads of sweat forming on his back.

Purdey looked a little fazed but quickly rallied. 'You didn't need to. Who else would be up there?'

Donald nodded gravely. 'Aye, we're infested with them too.'

Purdey's smile was confident now. 'They heard a shot, you say. That could have come from anywhere, couldn't it, boys?'

'Right enough, sir,' Hamish managed.

Donald smiled, nodded, and looked down at his hat.

Purdey rose from his chair. 'Impossible to say where that shot came from. My men were shooting hares, you see.'

Everyone stood up. The inspector folded away her book and gave the keepers a long, hard stare. Both of them found the office floorboards immensely interesting at that moment.

Purdey smiled and waved as the police car drove away and Hamish sighed with relief. Purdey clicked his heels. 'Told you they couldn't prove a thing. They didn't actually see you shoot anything.'

The keepers nodded.

Purdey turned to walk away, but then thought better of it and drew close to Hamish and Donald. He spoke quietly, as if they might be overheard – although the estate was deserted apart from the three of them.

'I just want to be sure, now. If the RSPB chap went wandering about on that moor, he wouldn't find any contaminated voles, would he?'

Donald shuffled. 'Aye, well, there might be a bag of them up there still.'

'Did you not go up and get them?'

Hamish risked speaking. 'Aye, we did but you see the moor's awful big ... '

'An' the voles is awful small,' Donald finished for him.

'I think you'd better have another look.' Purdey's tone was icy.

The laird marched off to the house.

The old folks' home was a modern red-brick building set back from the road close to the River Ness. The young woman who greeted them at reception had spiky, bleached-blond hair and a stud in her nose. She beamed at Rory when he rang the bell and explained why they were there.

'He's not had a visitor for nearly a year,' she said in perfect English but with a heavy Eastern European accent.

As they followed the young woman down the corridor, Rory caught the smell of disinfectant and overcooked, stodgy food – an aroma that took him back to visiting his gran in the care home in Birkenhead not long before she died. The young woman turned to them in the corridor beside a clinical-looking door with a card that read 'Mr Eoghan Campbell'. Written in blue biro beneath the word 'Eoghan' was 'EWEN'.

The nurse put her hand on the door and Rory noticed that her fingernails were varnished black. It struck him as odd that a nurse should have black fingernails, although he couldn't think why that might be.

'Sometimes he gets confused.' The nurse smiled and paused. 'Sometimes it's better to just go along with him. We'll see.'

She knocked on the door and opened it without waiting for a reply. 'Visitors, Mr Campbell,' she said cheerily, and ushered them into a small room that made Rory think of a hospital.

The old man, whose skin had the texture of crumpled brown paper, was sitting staring at a TV screen watching a show about selling antiques. He showed no sign of being aware of either of his two visitors. *How odd*, Rory thought, *that this man must have spent long nights sitting before the same bothy fire that we do, and wandering over the same hills as us.* All those days out in the wind and rain, and nights in that wild place, yet here he was, fading away like the image in an old photo.

The elderly shepherd did not react. His dim eyes stared into space.

'Visitors, Ewen,' she stroked his hand again.

Campbell turned this time. 'What is it, Sheilis?' Then he turned to Rory and Angus, as if noticing them for the first time, and said gruffly, 'Oh, you're here, are you?'

The nurse rearranged the pillows on the old man's chair. 'It's Svetlana, Ewen. Visitors.'

The old shepherd peered at them. 'Useless pair of fools. Bone idle, the pair of you. Are the peats in?'

Rory knelt beside the old man's chair. He didn't know how to begin to talk to the old man, and glanced up at Angus, who shrugged.

'Hello, Mr Campbell. We've come about the bothy.' The old man stared blankly at him. 'We're looking for the deeds to Glen bothy.'

The old man's jaw trembled. 'Deeds! You'll get no deeds from me. This is ma hoose.'

The shepherd turned away, lay back in his chair and closed his eyes.

The room was dimly lit now, the smoke from the peat fire hanging in wreaths around the rafters. Sheilis was bent over the fire. Steam rose from the kettle and white shirts hung dripping by the cast-iron mangle, its heavy wooden rollers squeezing the water from the newly washed clothes. He watched Sheilis pummelling the white linen clean. God, she looked old. Yesterday she had been tall and straight, her hair raven black; now she was bent, her locks streaked with grey.

Mr Campbell raised himself up in his chair. 'Where's ma dogs? It's time for the ewes. I cannae sit here.'

The two boys stood silently beside the shepherd.

'Go on, the pair of you. You get off to school and mind that schoolmaster now. You get certificates. You need certificates, or you'll spend your life tramping these hills for pennies.' He coughed, his frame wracked with each breath. 'I said, are the peats in, woman?'

Sheilis stood and smiled. 'Aye, the peats are in.'

Rory knelt at the old man's chair. 'You see, Purdey wants to take it back and we need the deeds.'

The name Purdey seemed to stir something in the old shepherd, and he began to fidget with long, spider-like fingers. 'Purdey, ha! Yon bastard. Well, he'll not get me oot o' this hoose. I have the papers!'

'Glen bothy,' Rory persisted. 'We wondered if you might know where the deeds are.'

The old man turned to Rory, his eyes suddenly animated. 'This is ma hoose. Yon young fella Muir gi' it me.' Then the old shepherd laughed so hard that Rory thought he might break his ribs. 'Hair like a girl. Yon hippy.'

Angus looked uncomfortable and began to fidget. 'I think we should go, yer ken.'

Rory sighed and decided to try one more time. 'You see, Purdey—'

'Bastard!' the old man yelled, and pulled Rory to him with surprising strength.

Svetlana stroked his hair. 'Now then, Ewen, don't get upset.'

The shepherd whispered to Rory, face so close that Rory could see the blood vessels in his eyes. 'Who's master o' this glen?'

'That's the thing. We're looking for the deeds,' Rory said.

The old man grew agitated, rocking in his chair. 'I cannae sit talkin'. There's ewes to bring in!'

'Let's go,' Angus whispered. 'He's nae all there, yer ken.'

Rory turned to the nurse and spoke to her quietly. 'I'm sorry we've upset him.'

She knelt and fussed about the old man, covering his legs with a blanket. As she stroked his wispy hair, the crumbling man eased back in his seat and became calm again.

'OK. Thanks very much, Mr Campbell.' Rory walked towards the door, feeling disappointed but sorry for the old shepherd, and Angus followed.

'Purdey can go to hell. Young Mr Muir told me it were mine.' The old man laughed again, his frail body shaking. 'Purdey signed it. The old place is mine.'

Then his eyes glazed and the old shepherd fell asleep, perhaps exhausted by the strain of having to communicate.

'There, he's sleeping,' the nurse said quietly. 'I've heard him mention Mr Purdey before. He does not like him one bit, I think.'

Angus chuckled. 'Aye, he's nae alone there.'

Rory was glad to get outside. The smell of the old folks' home still clung to him.

***

The rest of the week passed slowly, but Rory was pleased when Angus called him to arrange to visit Glen bothy at the weekend. Perhaps their climb had stirred Angus's old desire for the hills – or maybe the chance to talk about his accident with James had helped him. Maybe that secret had been eating away at Angus and now that it was out he could begin to find his feet in the hills again.

The air was cooler now. In early September, the trees were already showing the odd golden leaf as autumn began to creep in. Angus and Rory started slowly on the climb through the forest. They had walked through these woods in winter when the ground was frozen hard; they had come here in the spring when the woods had been bursting with new life. They had taken this path when the summer sun had blazed down on them and the air had hummed with insects and shimmering heat. They had watched as the landscape had settled beneath the crushed yellows and browns of the winter land; they had seen the green return in spring and burst into colour in summer.

'Autumn soon,' Rory observed, noting a handful of leaves that had lost their green.

Angus paused and took in the cool of the air. 'Aye, not long now.'

They could feel the mood of the place change as the year turned. Soon all the trees would blossom gold; soon frosts would touch the grass and glaze the puddles on the path. The first frost would take the insects out of the air, driving the clouds of midges from the glen until the next spring.

Only by returning to the same glen over and over can one learn to read the signs of time written on the landscape.

Of all the seasons, autumn arrives with the most drama. She comes in with a paintbrush and creates a sunset on the hillside, heralding the changing of the year. In autumn the trees draw their lifeblood back out of their leaves and deep into themselves as the wild things begin to brace for the winter. Creatures gather the stores they will need to get through the coming cold in a frantic race to survive. When the snows finally come, the weak and the old will sleep forever.

Rory stopped on the wooden bridge and rested for a few minutes, waiting for Angus to catch up. The bridge was new, a replacement for the one that had collapsed the night of the floods. The wooden rails were hard-edged and smelt of wood preservative. Rory ran his fingers along the rough surface. How long would it be before the frosts of winters to come, and the sun of future summers, bleached and aged the wood, softening its edges? He remembered the old shepherd sitting in his chair in the care home being comforted by the nurse. Mr Campbell must have walked across the old bridge a tall young man, and now the torrent of the years was washing away at him too.

Angus arrived at the bridge and dropped his pack, his breath misting in the air after the exertion of the climb. He pulled out a packet of chocolate raisins and offered some to Rory, who waved them away.

'Do you think the old shepherd was making any sense?' Rory said. 'The stuff about the master of the glen. Something about a stag. What do you think he meant?'

Angus shrugged. 'I dinnae ken. But I thought that he knew what he was talking about for just a few moments, and maybe there is something in what he said.'

Rory shouldered his pack. 'So let's search the bothy tomorrow. See what we can find.'

They crossed the bridge and followed the track out into open countryside, where it ran parallel to the river that had carried all three of them away only a few months ago. After a while the whine of a petrol engine rose above the gurgle of the river. They both stopped and turned in the direction of the sound.

'It's that bastard ghillie,' Rory hissed.

Donald was riding a quad bike towards them on the far bank of the river. He gave them an angry glare, which Angus and Rory returned with even greater belligerence; a moment later, Donald stopped his bike and turned off the engine.

'I had the polis asking me questions because of youse,' the old man yelled from the opposite bank.

'You shouldn't kill wildlife, then,' Rory called back.

Donald shook his head. 'I nivver did. Youse eejits keep your noses out of things that don't concern you.'

'We are within our rights if an offence has been committed,' Angus said, struggling to contain his temper.

Donald leapt back on to his quad bike. 'Ach, what would you know about it? An' anither thing—' The roar of the engine covered up his string of expletives.

'You've no right,' Rory yelled across the river, but the ghillie set off on the quad bike and Rory didn't bother finishing his sentence – Donald would never hear him.

The two hillwalkers stood silently watching the gamekeeper as his bike entered a thin line of trees beside the river. They saw the bike disappear, then there was a crunching metallic sound and the engine died.

Rory peered at the trees. 'What happened?'

They waited but there was no sign of the quad bike leaving the wood. The river gurgled on as Rory scanned the trees with his binoculars.

'Can yer see him?' Angus said.

Rory lowered the glasses. 'Maybe he's crashed.'

Angus cupped his hands around his mouth and yelled, 'Hello!' His voice sounded small amongst the hills, and he turned back to Rory. 'We'd better check he's OK.'

Rory nodded and began unlacing his boots. 'It's probably only knee-deep here.'

There is an unwritten law in the mountains: all rivers are deeper than they look. This river was no exception. In the first two steps the level passed their knees, and it cheerfully made it to their waists before they scrambled out of the cold water and headed for the trees.

It wasn't difficult to find the keeper. The quad bike was upside down, its wheels still turning, with the driver semi-conscious beneath it.

Angus grabbed the frame of the bike. 'Gi' us a hand here.'

Rory took hold of the bar and together they pulled the quad bike off the keeper. He rose from the ground like a Hammer horror vampire rising from the dead, a trickle of blood oozing from his head. Rory tried to support the keeper but the older man shrugged him off.

'Ah Jesus, that was some skelp that.' Then he felt his backside. 'Och, ma erse.' The keeper staggered about, clearly searching for something, then reached up and felt the wound on his head. 'Ma heid! I cannae find ma bunnet.'

Rory spotted the crumpled deerstalker in the heather and handed it to the keeper. Angus tried to help the man but was brushed off just as Rory had been.

The dazed keeper reached for his phone. 'Hamish, ye ginger eejit. Get yer erse up here – I've fell off ma feckin' bike, so I have.'

Within minutes the big red-haired ghillie was helping Donald into his Argocat, which was no easy feat as the injured man unleashed a ferocious barrage of obscenities every time he was touched. Hamish uttered a reluctant thanks and the pair roared away into the distance with Donald's swearing slowly fading from earshot.

Angus followed the progress of the two keepers with his gaze. 'Do yer ken he'll end up in hospital?'

'No,' Rory said. 'They'll probably just have him put down.'

AUTUMN
CHAPTER 10

Laura put the plate down in front of Angus and set about loading the dishwasher. Angus peered at the lumpy yellow conglomeration in front of him and prodded it wearily with his fork.

Laura gave a heavy sigh. 'Your scrambled egg makes such a mess of the pan.'

So that's what it is, Angus thought, and shovelled a forkful of the congealed mass into his mouth. Cleverly Laura had removed all the taste from the eggs, something she had a unique talent for.

The post lay unopened on the table and Angus began shuffling the envelopes. There were the usual adverts for takeaways and supermarkets promoting their latest special offers. There was even a glossy brochure from one of the town's outdoor shops. Angus leafed through the pages where slim young men and women grinned back at him in fashionable outdoor clothes that never seemed to get wet or dirty no matter what the weather. He mentally ran through all of his outdoor gear and decided, regretfully, that he couldn't justify replacing any of it for at least a year or so. He hurled the brochure into the bin. There was an envelope buried under all the advertising. His name and address were handwritten, a rare thing in these days of automated mailshots. It wasn't his birthday, and he couldn't think of any friends who might be getting married, but the envelope had that special-occasion look about it.

He slid his breakfast knife down the side of the envelope, and a card emerged, covered with party balloons.

He read the invitation out loud: 'You are cordially invited to a ceilidh celebration at the residence of Mr Tony Muir.'

He expected Laura to dismiss the idea, as she always did when he suggested she join him on his adventures in the wilder places of the Highlands.

'An invitation from Lord Muir,' she murmured. 'We must go.'

Angus was puzzled. 'But yer dinnae like staying in bothies.'

Laura sniffed. 'We won't be staying in any *bothy*.'

Angus felt a cloud of depression fall over him as he realised that Laura would never turn down an invitation by someone she saw as landed gentry.

'He's nae a lord. His da made a fortune selling biscuits, yer ken. And he won Glen bothy frae Purdey in a card game years back.'

'And I am not,' Laura continued as if he had not spoken, drawing herself up to her full nearly five feet, 'attending a social occasion at the Muir estate house arriving in a battered old Land Rover like some farmer's wife.'

'There's nothing wrang with being a farmer's wife,' Angus allowed himself, but he knew it was futile. Laura had already made up her mind.

Angus had never been a social animal. He liked a quiet bothy night or perhaps an evening with a few friends in the local pub. There would be noise, and lots of people he didn't know or want to know all thrust together, but worse than that – *far* worse than that – there would be dancing. Angus hated dancing with a passion. He could never understand the connection people felt between movement and music, although he knew that once Laura had set her mind on something there was no way of changing it. So they would be going to the ceilidh.

***

For the first time in his life, Angus was setting foot on soggy Morvan dressed in a suit. He often wore suits for work and the occasional civic function but here it felt out of place and uncomfortable. He turned Laura's blue Nissan through the wrought-iron gates of the Muir estate, and followed the long gravel driveway as it wound between the rhododendron bushes. It was early evening by the time they arrived at the car park, which was already full of the cars of locals.

Rory had fallen asleep in the back of the car. Jen had to prod him to get him to wake up, and he sat up, rubbing his eyes.

'I've never been to the big house before. We always just go to the office.'

Tony Muir was standing at the entrance of his impressive house dressed in Bermuda shorts and a Hawaiian shirt. His only concession to the ceremony of the evening was that he had, for once, replaced his battered wellies with sandals. This left him with the appearance of wearing long white socks as his shins were so rarely exposed to sunlight that they contrasted markedly with his sunburnt knees. It's hard to say if Tony Muir was aware of this wardrobe malfunction, but he certainly wouldn't have cared even if he knew.

'Is that him?' Laura asked, looking a little surprised at Muir's casual appearance.

Rory was easing himself out of the car. He was less formally dressed than Angus, but Jen had managed to get him to put on a clean, ironed shirt – a step up from his habitual T-shirt and jeans.

'Yes, that's Tony.'

Angus walked round to Laura's side of the car and as she took his arm he realised that it had been a long time since they had walked anywhere like this. Years ago it was something they had done all the time.

As the little group of hillwalkers headed towards him Tony beamed his welcome.

'Ah, the bothy dwellers!' he cried.

Looking at his wife's face, Angus could tell that Laura was wincing inside, even though she did an amazing job of not showing it. She didn't actually curtsy, but it probably crossed her mind.

Tony shook Angus's hand warmly. 'You're looking smart tonight, man.'

Angus felt uncomfortable in his grey suit, especially next to Tony and his peacock display. 'I thought I'd better make an effort.'

Tony opened his hands, showing off his own colourful if shabby outfit. 'Putting me to shame, Angus.' He gave them all one of his extra-wide grins. 'Still, I am disgustingly wealthy, you see, which means I never have to give a damn what people think. It's a perk of the job. You'd better all come in and have a drink.'

Inside the house, the great hall had been decked with flags and bunting. Down one side of the room there were trestle tables where an informal bar had been erected, with bottles of beer, wine and whisky glittering in the light of the candelabra. On the other side of the hall there were similar trestle tables piled high with sandwiches. Angus noticed an improvised stage at the far end of the hall, with four musicians struggling to find space for themselves and their fiddles and accordions. Everyone Angus had ever seen in the village was there. Around fifty folk milled around. Christine from the little grocery store, he noticed, was busily engaged in conversation with the young Polish woman who worked at the local fish farm.

Sinead, who had organised the whole thing, came over to the walkers. She had an air of efficient organisation. Her dark hair was tied back and she looked classy and elegant, her scruffy jeans and fleece replaced by a red dress.

'Hello, Sinead. I almost didn't recognise you,' Rory said, beaming, and he meant it – the transformation was almost as astonishing as Angus's. 'You must be the woman in red.'

Jen grinned from beside him and feigned offence. 'Nah then, tha'll be in trouble if there's any more o' that.'

Sinead laughed, and she might have blushed a little too. 'I've something I need to say to you two.'

She drew Rory and Angus slightly to one side just as the band started to play a reel, and the MC, a stout man with a Father-Christmas beard, started to call on folk from the crowd to join together. A few stepped forward. Those who had drunk enough to take part in the dancing listened to the MC's complex instructions about whose hand to hold and which side to pass which person.

Sinead glanced over to the dance floor and, happy all was going to plan, turned back to Rory and Angus. 'I've been asked to have a word with youse two fellas,' she whispered. 'Look thonder in the far corner.'

Over in the corner sat two men – one with a bandage around his head; the other younger, bigger, with a shock of ginger hair. They were dressed in tartan kilts and formal jackets. Out of context, indoors and not wearing their everyday tweed uniform, it took Rory a moment to recognise them as the two keepers who had shot the harrier. They were sitting upright, stone-faced in their kilts, and obviously feeling out of place.

Rory felt his stomach churn. 'It's those two bastards!'

'They've some nerve,' Angus said.

'Now, then,' Sinead said firmly, 'they want to thank you for what you did for Donald the other day. They think they owe you that, but they're a bit scunnered cos they don't know how youse'll take it.'

'I'll give them a bloody reception,' Rory said loudly, ignoring Jen's uneasy looks from beside him.

Sinead pushed his hand down. 'It took me hours to organise this here hooley. I don't want you spoiling it.'

'I don't want their thanks.' Rory turned. How could he forgive them for killing the harrier?

Jen took Rory's arm. 'Look, 'appen they's trying to do the right thing. Don't you go belt one of 'em an' show us all up, you berk.'

Rory's eyes flamed again and he turned to Jen. 'You didn't see the hen harrier.'

Sinead put her hand on Rory's chest and steadied him. 'It's nae really them. They has to make a living, so they do. It's Purdey you should be blaming.'

Angus put his hand on Rory's shoulder. 'Dinnae worry, Rory. There's nae point in holding this against them. Sinead's richt.'

Rory reluctantly nodded and Jen stroked his shoulders.

The keepers stood before them, fidgeting in their Sunday kilts.

Donald addressed Angus, perhaps because they were the two most senior present. 'Jis' wanted to say, I'm grateful for what you did for me when I came aff the bike.'

There was an awkward silence and then Donald proffered a bottle of whisky, which Angus took reluctantly. On the west coast of Scotland there are two types of currency. There are pounds sterling, in which people are paid, and with which houses are bought and sold and everyday goods change hands. Then there is a second currency, whisky, which is used for many things that cannot be done with ordinary money. There is not a locked gate on any track in the Highlands that cannot be opened by a bottle of whisky placed in the right hands. Deals are done with bottles, apologies made and arguments settled.

Angus smiled. 'Aye well. Couldnae leave you for deid, yer ken.'

Hamish looked at the floor but held out another bottle in his big paw for Rory.

Rory held back but Jen pushed him forward and he slowly reached out and took the bottle. 'We don't think much of what you did to that harrier.'

Donald answered for both of them. 'Aye, I cannae say I'm proud.'

'Yer ken what yer did, then?' Angus asked, surprised at the older man's frankness.

Hamish cut in. 'You come here for fun. We've a living to make. And there's a big shoot tomorrow.'

'You are destroying the environment, you know that?' Rory said, unable to contain himself.

Hamish took a step towards Rory. 'What do you know? It's good for the land. What we do maintains a diverse environment.'

'That's crap, you know that.' Rory was staring Hamish in the face now.

Sinead stepped between them and spoke forcefully. 'Don't spoil my feckin' ceilidh now, boys. Let's have a wee truce.'

The two of them were staring hard at each at each other, both about to spring.

Donald pulled Hamish back. 'Haud on, man. There's nae need for this.'

Rory stepped back too, and the tension eased.

Donald touched the bandage on his head. 'Thanks, man.'

Angus nodded, and held up the whisky bottle by way of thanks. 'Fit's yer heid?'

The older ghillie smiled and rubbed his backside. 'Ah well, me heid's fine – it's ma bahookie that's bruised.'

The two older men laughed, then Hamish muttered some-thing that may have been an apology and the keepers walked away through the dancing folk.

Rory looked down at the bottle in his hand. 'We shouldn't have taken it.'

'They meant well, tha knows,' Jen said to Rory, still holding on to his arm as if she thought he was about to bolt after the ghillies and use the bottle as a weapon.

Tony Muir appeared. 'Everything all right?'

Sinead was quick to reassure him. 'Aye, it was jis' Purdey's ghillies saying thanks.'

Tony relaxed. 'I heard about that business with the quad bike. They can be dangerous, those things – good job you showed up.'

Rory saw Angus catch Tony Muir's eye. 'I was wondering if you had got anywhere finding the deeds to the land,' Angus said.

Sinead shook her head and answered before Tony could reply. 'Nah, we've looked everywhere, so we have.'

'Rory and I went to see Ewen Campbell, the auld shepherd who lived in the bothy years and years ago,' Angus explained to Tony. 'But he's pretty senile now, yer ken, poor soul.'

Rory was trying to pick out a vegan option from a selection of sandwiches. 'He did say something about the old bothy. Something about a stag being the lord of the glen.'

Tony's eyes brightened. 'The bothy, eh? Worth a look there, do you reckon?'

Angus and Rory exchanged glances. They'd both had the same hunch.

It was settled that Angus and Rory would search the bothy tomorrow. Neither was optimistic that they would find anything – there could be few hiding places in the sparsely furnished bothy – but there was always the chance that the deeds were in some hiding place that no one had ever seen.

***

Jen was having a good time. As the evening continued and the band grew louder, the dancers struggled to follow the MC's instructions. They formed lines and twirled about each other. Some had performed these dances since childhood and followed them effortlessly. Others turned the wrong way or barged into people and had to be pushed in the right direction.

Jen and Rory took part in a few of the dances, and then sat down, breathless, at the side of the room.

Rory took a sip of beer and laughed as he watched the people milling about the dance floor. 'Sometimes I think Highland dancing is less of a dance and more of a martial art.'

'Tha could be reight there,' Jen said, struggling to make herself heard over the music.

She noticed that Rory was looking up at an old picture hanging on the wall. It was a romantic oil painting of the Isle of Morvan. Jen realised that he was dreaming again of moving to his beloved island. She worried that the reality might be very different to Rory's vision. He saw a quiet, remote place with long sunny days where he could hide away from the world, grow vegetables and become self-sufficient. She saw weeks of rain and wind and the end of her nursing career. Then she saw a brown envelope, a solicitor's letter and a cheque for 80,000 pounds.

Rory turned to her. 'I've been thinking. If I worked some weekends maybe we could save more.'

Jen looked up at him over her wine glass. She knew she should tell him about their windfall, but if she did then he would want to move somewhere instantly.

'But tha likes going t'hills at weekend.'

Rory shrugged. 'I know, but maybe for a year or so … and then, if we saved enough money, we could buy somewhere.'

She could see he was growing desperate. He had to be to want to spend more time in BetterLife. 'Maybe it'll take a while, tha knows.'

He turned to her with a pained look in his eyes. 'Only way I can think of, unless you have a better idea?'

Jen felt the words bubbling up in her throat. She wanted so much to tell him about the money – she knew she had to, but perhaps not yet. She didn't know when the right time would be but couldn't bring herself to say anything right then. He was agonising about how to get a new start, and she held the key in her hand but couldn't find the courage to give it to him. *What if I really can't move to his island, what then?*

'We'll find a way, tha knows,' she said quietly.

At that moment Tony Muir appeared beside her and asked her to dance with him. She was glad to get away from that question, even if it was only for a few minutes of losing herself in the swirling bodies.

***

As the night drew on, after a few more whiskies than was good for him, Angus went outside to the veranda of the Victorian house to smoke his pipe. The air was cool and the night was still. A big moon peeped over the hilltops and shone down on the lawns through the trees. He packed his pipe with Irish Cask tobacco and lit it, revelling in the peace and quiet away from the band. Angus had never liked the music of the accordion. It sounded to him as though accordion players only ever knew one tune. They simply played it at different speeds on different days. As the pipe smoke soothed him, he spotted two figures walking on the lawn. The first he recognised instantly as the diminutive figure of Laura. The second figure was clearly a man but at first Angus couldn't make out who it was. He stepped down between the stone columns at the edge of the veranda to get a better look. The pair came out from the shadow of the trees and, just for a fleeting moment, Angus glimpsed the thin figure of Councillor Gerald McCormack talking intently with his wife. They were not touching, not quite, but there was no mistaking the intimacy between them.

When Laura turned to walk away, McCormack said something Angus couldn't hear and then Laura turned back to him and reached out and touched his cheek. It was just a moment, just a fleeting touch, but Angus knew it for what it was: a lover's touch. Then the pair separated on the lawn and Laura began making her way back to the hall.

Angus drifted back into the shadows. He stayed outside smoking his pipe, and as he did so, he began to remember things. Things like the two wine glasses in the dishwasher, like how keen Laura always was for him to go away at weekends – she had even pushed him to go on that climb with Rory. Over the past couple of years he'd felt that something odd was happening. Laura had changed ever so slightly in ways he couldn't quite define. Now he knew with absolute certainty what had been going on in front of his eyes. Part of him couldn't understand why he'd not seen it before – perhaps he'd not wanted to – and part of him just felt numb.

Angus was lost in thought when he heard a woman's voice behind him. 'I couldn't trouble you for a light, could I?'

He turned and found Tabatha Purdey standing behind him with an unlit cigarette in her hand. 'Aye, nae problem.'

Angus handed her his matches and she smiled at him.

'Thank you.' She lit her cigarette and inhaled deeply. 'I'm Tabatha Purdey. I saw you in the village shop, didn't I?'

Angus squirmed, praying she didn't remember him from the incident on the riverbank. 'Aye, yer did.'

Tabatha stared at Angus. 'I just can't get over the feeling I've seen you somewhere before.'

Angus shook his head as Tabatha returned his matches. 'I dinnae ken.' He smiled and turned to walk back to the party, not wanting Tabatha to be able to see him too closely.

Before he'd gone two steps Tabatha called to him. 'Ah, I have it.'

Angus turned back, hoping she didn't have it. 'Aye?'

'You're one of the *naked* men,' she said, beaming in triumph.

Angus couldn't think of any other tactic so decided to deny it. 'Er nae, I dinnae think—'

Tabatha dismissed his argument before he'd even had a chance to make it. 'Oh yes. I never forget a face. Or anything else for that matter.'

Angus blushed.

Tabatha was in full flow now. 'Black magic, was it? A sort of ritual? Don't worry – that sort of thing has been going on for ages around here.' Then she hurled her cigarette butt into the shrubbery. 'Your secret is safe with me. We could do with a bit of black magic in these parts, don't you think?'

With that Tabatha marched back into the party, leaving Angus speechless in her wake.

At last midnight arrived and many of the guests were tired from dancing or sleepy with alcohol. After hours announcing dance after dance, the MC, still dapper in his dark suit, drew himself up to the microphone.

'And now, ladies and gentlemen, we have a special treat: Donald will play a lament on the pipes.'

There was a stirring in the audience. A few people rose; Hamish slunk out of a side door. Donald walked on to the stage, his head still bandaged from his encounter with the tree. He raised the pipes to his lips, closed his eyes and began to play.

If you had been standing outside the hall at that moment you might have guessed from the mass exodus of people that a fire had broken out. People streamed from the building while Donald played on, eyes closed, oblivious. A tortured sound came from inside the hall. It was as though the gates of hell had been opened and all the souls within cried out their torment. Donald played on, eyes shut, concentrating. When at last he took the chanter from his lips and opened his eyes he was alone in the hall.

***

The next morning, Laura and Jen sat in the passenger seats of McCormack's huge black Mercedes four-wheel drive. Laura could barely see over the dashboard, but Angus thought that she looked proud of herself up there. Angus and Rory – back in their scruffy hill clothes – stood with their rucksacks on, waiting to walk in to the bothy, watching the big vehicle throbbing with the diesel engine.

Laura leaned across the imposing bulk of Gerald McCormack and called to Angus. 'It's very kind of Gerald to give us a lift home, isn't it, Angus?'

Angus looked deep into the eyes of Gerald McCormack, who suddenly began staring intently at his dashboard. 'Indeed it is. Very kind o' him.'

They watched the all-terrain vehicle head off up the narrow road, punching a bigger hole in the ozone layer with every mile it travelled.

Rory stood watching the vehicle recede into the distance. 'I wonder how small a penis you have to have before you need to drive something as big as that?'

Angus watched the vehicle until it disappeared, trying not to think about what was going to happen when Laura got home. He turned and set off up the track. 'Aye, I wonder.'

Usually the road end was quiet, but today the start of the track was guarded by police and smart-looking young men who were trying so hard to blend in with the scenery that they stood out.

A police sergeant placed himself on the track in front of the two walkers and smiled. 'Are you heading to the hills?'

'We are,' Rory replied.

The police sergeant smiled again but Angus was pretty sure he wasn't going to say anything funny. 'Thing is, there is a shoot going on and for your own safety this road is closed.'

Rory looked annoyed. 'Right to roam legislation says we can go where we like.'

The policeman smiled again. It was one of those smiles that policemen use instead of riot shields. 'Of course it does, sir. If you'd like to come this way I'll show you the alternative route.'

Angus looked around the car park. There were lots of expensive-looking four-wheel-drive vehicles and a dozen or so men in new Barbour jackets trying to keep a low profile. There was something military about these men. Then he noticed that all of them wore earpieces.

'Is there some sort of royalty here?' Angus asked.

'I couldn't discuss that, sir. Now if you'd like to come this way ... '

The policeman held out his arm, and when neither Angus nor Rory moved, the men with earpieces started taking an interest in them.

For a moment Angus thought that Rory was going to argue again, but then he just shrugged as if it simply wasn't worth wasting any more time.

'Oh, come on, then.'

They walked off in the direction the policeman had indicated, towards a place where a section of fence had been taken down to allow access to the field. The police sergeant escorted them a few hundred yards and only left them when they were going the right way. At first they were in an open field; then the only route lay across a steep hillside. There was no path, so they spent their time stumbling through deep heather and cursing as the midges began pestering them with their incessant pinprick bites.

Angus stopped and surveyed the heather, blowing hard from the climb over the rough ground. 'Bad place for ticks, this.'

'I don't mind the bites, it's the risk of Lyme disease that worries me.'

Angus shook his head sadly. 'Aye, it's strange, yer ken. Years ago I'd heard of Lyme disease but I never knew anybody who had it. Now I ken lots of folk.'

Just talking about the little parasitic creatures had made him itch with imaginary bites. They moved on up the hill towards the edge of the Purdey estate and a place where they might be able to rejoin their usual route to the bothy. Now they could look down on to the wide valley below them where the driven grouse shoot was getting underway. A line of men, maybe twenty or so, were standing behind low walls with several shotguns each, and assistants with them to help them load. In the distance men with flags and whistles were fanning out across the head of the glen.

***

Hamish was at the centre of the line. 'Slow and steady, boys.'

Donald was a hundred yards away with his flag, and called to Hamish. 'What dae ye think?'

Hamish surveyed the ground between the guns and the line of beaters he was standing in. Already, small brown birds – the grouse that were their quarry – were bobbing up out of the heather or taking short flights on their stubby wings, fleeing from the line of men. Hamish tried to estimate how many there were. *Are there enough to make the shoot a success?* All their year's work had been building towards today, and the thought that there might not be enough birds turned Hamish's stomach. They had shot hares, trapped stoats and weasels, poisoned foxes and shot the hen harrier, yet still the vermin had come to feast upon their grouse. Had the keepers done enough? The next few hours would decide the answer to that question.

The radio in Hamish's pocket crackled into life. 'Are you there, Hamish?' Lord Purdey asked.

'Aye, sir, I'm hearing you.'

The radio hissed and squawked. 'Are your chaps ready at your end?'

Hamish looked to his left and his right. All the beaters were in place. 'Aye sir. We are ready.'

'Good. Bring them on, then.'

Hamish raised his flag and blew a whistle. Everyone in the

line turned to look at him. He dropped the flag and the men began to advance, yelling and waving their flags. Birds started to emerge from the heather in front of them. Hamish turned and Donald was grinning.

The line of men began to advance towards the shooters. At first only a few birds showed themselves, but soon there was a cloud of fluttering wings dancing in front of the line of beaters driven on by the flags and the yells. Then the guns opened fire. Each of the men in the shooting butts had two others to load his guns. They fired both barrels, handed the discharged gun to one man, and took a loaded weapon from the other. The firing went on continuously. The grouse fell from the sky in their hundreds. The shooters and loaders worked in a well-rehearsed rhythm, loading and firing without pause.

Hamish was happy as he marched forward with the line of beaters. The air was thick with grouse and the glen was filled with the pop, pop of shotguns as the men in tweed mowed down the birds. In all the noise he couldn't make himself heard, but he waved over at Donald, smiling. The old ghillie must know the day would be a success now, and he smiled back at Hamish. Their year's work, all the trapping of stoats, weasels and foxes, the hundreds of hares shot, the hours out in the rain and the wind, coming home cold and exhausted, had been worth it.

The grouse fell in the heather by their hundreds, feathered targets, dying for the pleasure of rich men.

***

High on the hill, Angus and Rory sat watching the shoot below them. Angus scratched his beard. It was like watching an alien army fighting a bizarre battle where the odds were stacked against one side.

Rory was studying the shooters through his field glasses, scanning along the line of butts. 'What do you think they get out of doing that?'

'I cannae tell yer. It's a mystery to me. They must like killing things, yer ken.'

Rory shook his head as he watched the carnage below. 'To think there are hundreds of square miles of Scotland's hills turned into grouse moor like this.'

Angus had never thought how much land was occupied by grouse moors. 'As much as that?'

Rory nodded, getting up from the heather. 'Aye, maybe as much as eighteen per cent of Scotland. No one really knows.'

'What a waste,' Angus said, shaking his head in amazement.

Rory took one last look through his binoculars. 'Bloody hell! Take a look at that bloke next to Purdey.'

Angus took the binoculars and turned the knurled knob until the man came into focus. He looked somehow familiar. 'I reckon yon's the heir to the throne.'

Rory shouldered his pack. 'Aye, that's who we're up against. The most powerful people in the country.'

The air was thick with the smell of gun smoke and blood. Still the birds came and still they were shot.

Angus lowered the glasses. 'Let's get awa' from this bloody din.'

As they walked away up the hill, Rory called to Angus. 'I think they do it because they can. It's about class – it's about people who own land saying to everyone else that they can do as they please. They do it because they can and because they have done it for years and no one can stop them. Prince Charles! That shows you who we're up against.'

The sound of firing followed them all the way up on to the ridge; they could still hear it when they reached the bothy. When they pushed open the rough door the little dwelling was empty, but a cluster of empty beer and cider cans lay scattered beside the dark fireplace as evidence of the most recent occupants.

Angus started dropping the cans into a plastic bag, seething with frustration at finding litter in the bothy. 'Wa's wrang wi' folk? Why can they nae take their junk hame?'

Rory just sighed. 'If it's open to the public you'll always get some who'll abuse it.'

Angus threw the last of the cans rattling into the plastic bag. He stood and surveyed the bothy, wondering where on earth to begin looking for the deeds. In truth there were not many places to look. The simple stone dwelling had only four rooms – two downstairs and two up – and precious little in the way of furniture. There was a table and a few chairs and an old wooden bench. But despite this, something well hidden could be in any of a thousand places. It could be in the roof space, hidden beneath floorboards, or tucked away behind the wood panels on the walls.

They put on their head torches and crawled into the loft space, coughing and choking on the generations of dust they found there. They took up floorboards and peered underneath but found no sign of Purdey's napkin – although they did succeed in unearthing copies of the *Daily Mail* from the 1970s, telling the same lies it still does today.

Rory wrenched up one of the floorboards and held up a small brass-coloured many-sided coin. 'What's that?' he asked, passing it to Angus.

Angus held it up to the light. 'Aye, that's an auld thrupenny piece.' He turned it over in his hand and read the date on the back. '1968. Must have been dropped by some auld hippy years ago, yer ken.'

Rory examined the coin. 'Makes you wonder, doesn't it, how many folk have passed through this old bothy over the years.'

By the evening they were spent, filthy and dusty from crawling around the bothy – and dispirited too, having found nothing. It was still light and warm enough to sit without a fire, but they put firelighters and a few sticks of kindling into the hearth and lit it anyway.

There is more to fire than heat. We have an ancient bond with the flames. Ten thousand years ago, people sat before a fire in the darkness. It made them feel safe and the light pushed back against the unknown beyond. Fire raised us above the beasts and provided us with a way of serving even the coldest night. Even now, where there are no beasts to terrorise us in the darkness, we are drawn to the power of fire, fascinated by the god in the flames.

Angus drew one of the whisky bottles the ghillies had given them from his pack. 'I think we've earned this today.' He read the label. 'Ardmore, ten years old.'

Rory was a beer drinker who rarely touched spirits. 'Is it good whisky?'

Angus poured a large measure into his mug and savoured the aroma. 'Aye. There's nothing like the smell of a good whisky in a bothy. Will yer nae try some?'

'OK, let me see what the fuss is about.' Rory held out his mug and Angus poured some of the golden liquid for him.

Angus savoured the first sip. 'Ah, you've never really drunk whisky until you've had it in a bothy on a dreich day, yer ken. It's like a good meal – nothing tastes better than when yer really hungry.'

Rory laughed. 'Well, there have been a few drams enjoyed in this place, that's for sure.'

Angus nosed the whisky in his glass. 'This is a braw drop. I ken coming here for the first time. There was a blizzard – God, the weather was wild. Must be thirty years ago. We got lost, spent hours looking for the place until we ken the dim glow of a candle in the window. We were trachled when we got here. An old timer inside, my age now, I suppose, handed me a bottle o' whisky and told me to take a good swig. It was like drinking liquid warmth. I could feel it glowing in my stomach and then it ran out into my fingertips. Five minutes later I felt fine.'

Rory tried his glass. 'Oh, I could get used to this, but I think I'd better not. I'll stick to beer.'

Angus filled his pipe and they sat in silence watching the flames dancing up into the chimney.

There was a knock at the door. Angus looked at Rory, puzzled; people don't usually knock on bothy doors.

'Aye,' Angus yelled.

It swung open and a tall, tweed-clad figure entered. It was Donald, the bandage on his head covered by a deerstalker.

No one spoke. He stood in the doorway looking awkward, like an interloper from another world. 'I'm here tae tell youse something.' As he spoke the small dark room filled with tension.

Angus broke the silence. 'We were jis' drinking your whisky. It's good, yer ken.'

The ghillie stood awkwardly and removed his hat, as one would stepping into a house to show respect. As he walked into the bothy he reminded Angus of a gunfighter entering a western saloon. But Donald didn't look as though he had come for a gunfight.

The rules of hospitality were written deep in Angus. 'You'll take a dram?'

Donald relaxed a little. 'Aye, I will.'

He took a seat by the fire and Angus handed him the only other drinking vessel in the bothy, a chipped old mug with a flower pattern on it. Donald didn't seem to notice. 'Thank you.'

Rory spoke reluctantly, as if treating the hen harrier killer with any kind of respect was treachery. 'We thought we saw someone on the shoot today.'

'Aye, it were him – Charlie was there,' Donald said with a nod. 'He's up at the big house now, having some scran.'

Donald sat looking into his whisky, as if searching for something to say but the words weren't forming. At last he found his voice. 'Sinead told me youse are looking for deeds?'

Rory put down his glass. Angus could tell that the younger man didn't want to drink with the old keeper.

'We have been searching for them all day,' Rory said. 'I hear Purdey means to close this place if we can't find the deeds for Tony Muir.'

Donald nodded slowly. 'That's about the size of it.' He lapsed into silence again, lost in thought now. 'Youse maybe think

I've a brass neck coming oot here after what happened.'

Rory's anger was rising again. 'How could you shoot that harrier?'

The keeper put down his cup and sighed, and Angus saw for the first time that the man was troubled. The ghillie rose to his feet.

'Aye, well let me show ye something.' He led them out of the bothy and a few hundred yards to where the crumbled remains of a stone wall rose out of the heather.

The keeper bent and touched the wall with great gentleness. 'This glen is empty naw, but it was nae always so.' His voice shook a little and the hillwalkers listened to him in silence. 'See this hoose here by the way? This was ma great-grand-da's hoose. There was eight or nine hooses here. Ma family had lived here as long as anyone could mind.'

He laughed, and Angus thought he was trying to hide his sadness. 'They wis poor, ma family. Life was hard, but this was our place.'

He looked up at the hills crowding in on the steep-sided valley, watching as the clouds swept the summits. The old ghillie could see a long way, Angus thought; his eyes took in a thousand years, a hundred generations.

'Trouble wis, the land wisnae ours, no on paper anyway. There wis money in sheep in them days.' He laughed, now that he had found his voice words tumbled out of him. 'No like the day. They wis cleared from here by the landlord, put off the land like so many cattle. No jis' this glen but the folk over the hill, on and on fae every glen.' He was on the brink of tears when he stopped speaking to run his fingers over the rough surface of the stone once more. 'If these stanes could speak they would greet. Aye, and the name of the man who moved ma family off this land wis Purdey.'

***

Back in the bothy, Donald sat sipping at his whisky from the chipped cup and Angus listened as the ghillie told of the Clearances. He realised that he was listening to a man whose heart was still full of the place.

'They burned the thatches of the hooses so as they couldnae come back,' Donald was saying. 'They say the air was honkin' wi' smoke for miles around. Every glen, every township.'

Rory had lost all his anger. 'What happened to them, the folk who lived here?'

Angus filled Donald's cup with whisky and the keeper sighed.

'They sent them tae the coast. They wis supposed to take up fishing. What did they know aboot the sea? Most of 'em had lived their whole lives here. They just scattered. Dust across the world. Some to Canada, others Australia. Ma family went to Glasgow. They wis farmers, what could they do there?'

'But you came back?' Angus asked quietly.

'I never meant to stay, you know. I jis' wanted to see this place fae massehl. The place I came fae.' Donald spoke as though he were speaking to himself, unaware of any audience. 'Oh, the rain. I mind it pished doon. The burns wis fit tae burst. But I came here anyway, an' I stood amang the ruins. I knew right away I couldnae leave. I wis home.'

Rory let the keeper speak, and then asked, 'How do you manage to work for Purdey if you feel like that?'

Donald shrugged. 'How else am I to keep close tae this glen? I dinnae *enjoy* it, but what can I do? I'm scunnered by what we do here. Once this glen wis home to folk. Once there wis crops, cattle wis raised here, weans were born and the land lived. If Purdey gets hold o' this place he'll turn it into a killing desert jis' like the rest of his estate.' Donald hung his head. 'An' I know I play my part in that. I make nae apology, but I can try tae stop things getting any worse.'

Rory looked amazed, as if he found it difficult to believe that he and the keeper had so much common ground. 'I get so angry,' he said. 'Scotland is unique – nowhere else in the world is ownership of the land concentrated in the hands of so few people. And they treat it with contempt. Destroy its wildlife, bulldoze tracks all over it. To think so little of the land … and for what? For a few days' shooting a year.'

'I dinnae ken why they do it,' Angus said as he leaned forward and pitched another handful of twigs on the dying fire. 'Such a waste, just for a few grouse or for some rich German to shoot a few stags. It must be the money, I suppose.'

Donald shook his head. 'I used to think like that. I thought it was for the money but that's nae it. They don't need money.'

'Why, then, do what they do to the land?' Rory asked.

'I wis driving Lord Purdey across the moor, and he wis on the mobile talking to one of his politician friends about a bill the Green Party were putting forward tae make hill tracks subject to planning permission. An' he says to this politician, "I'm no putting up with having to go to the council to ask if I can build a track on my own land. I'm going to damn well do what I want and the more they try and stop me the more I'll do it to show them who's boss.'"

Rory nodded. 'They do it because they can. It's to show people like us that they are in charge.'

Angus listened to Rory talking to the old ghillie. Not that long ago Angus would have argued for the status quo, but things he had seen over the past few months made him feel that perhaps there had to be a change in the Scottish countryside. Nevertheless, something in him still felt the need to argue for balance.

'But it's nae all their own way, yer ken. Look at access rights. The right to roam.'

Rory was growing angry again. 'Yes, we can go where we like. We can walk across a bloodstained desert and call it freedom.'

'Youse'll never stop them,' Donald explained slowly. 'Youse are up against the most powerful folk in the land.'

Donald talked more than Angus had ever heard him talk before – like a monk who, amongst people who did not follow his god, could suddenly rail against the sacred. He was speaking blasphemy, but he was amongst those who understood the need for sacrilege.

It was late – and the bottle of Ardmore was almost empty – as Donald made his way a little unsteadily to the door. He stopped and turned to the two walkers, his hand on the doorknob. 'Och, haud on, I nearly forgot. The whole reason I came. Campbell told ye the deeds are in the auld bothy, did he naw?'

Angus had no idea how Donald could know that they'd talked with the old shepherd, but he had long since ceased wondering how news travelled so effectively around the village. 'Actually it were more of a hunch. A wild goose chase, yer might say.'

Donald brushed that aside. 'The auld bothy, he telled ye? Youse'll nae find the deeds here, then.'

Rory looked at the keeper, surprised. 'Why not?'

Donald took a deep breath. 'This is nae the bothy Campbell means. I mind hearing of another bothy, swallowed by the forest years ago. *That's* the auld bothy he's talking aboot.'

Angus looked at the old keeper with a renewed respect. Donald had just betrayed his friends and his employer, and no matter how little regard he had for Purdey, that said something.

'Where is it?' Rory asked.

Donald crammed his deerstalker on to his head. 'There's the problem right enough. It's been lost for years, but if ye follow the river into the forest ye might find it. If the deeds are in the auld bothy, that's the place to look.'

CHAPTER 11

The sky was dark and a cold rain was falling steadily as Angus stepped out of his car and walked towards the chapel. Laura took his arm and Angus was glad that she was here to support him now. Over the years their marriage had lost its passion. The young woman he had gone climbing with on Skye had been full of sparkle, but time had taken that away. At times like these, when he was at his darkest, Laura was still there for him. As she steered him towards the chapel door, somehow, he knew, she sensed the dread he was feeling and knew that he needed her small, strong presence.

Rory was waiting beside the door of the crematorium. He'd squeezed into a dark suit; Jen, neat and black, stood beside him. Angus noticed that large raindrops jewelled Rory's ginger beard and wondered how long he had been standing in silence outside the low, red-brick building. Rory looked uncomfortable in his suit. He reminded Angus of the chimps that used to dress up and have tea parties to entertain children in the zoo. You couldn't trust them because even though they looked quite civilised you knew that at any moment they were likely to pee in the teapot and hurl excrement at the nearest child.

Angus felt the strength leave his legs as he walked toward the chapel, and he hesitated in the doorway. Rory was beside him instantly and he felt an arm around his shoulder.

'Come on, Angus. Let's get this over with.'

The two couples walked in out of the rain and Angus was surprised to see the number of folk in the chapel. About a dozen of the club members were there; Gary turned to smile at him. There was a tall dark middle-aged man sitting awkwardly on the front bench. His head was shaven and Angus could just see the blue streaks of tattoos on his neck – this must be the son he had heard the police talk about. Chris and two other members of the rescue team sat towards the back. Angus was surprised to see six other men lined respectfully at the back of the chapel, each with a neat maroon beret folded on their lap. On a second glance Angus saw the insignia of the Parachute Regiment on the caps.

There was music playing quietly in the chapel, but it ceased as a large bearded man entered. He was sombrely dressed in a black suit. 'Ladies and gentlemen, my name is Phadrig and we are here to celebrate the life of Brian Murphy. Brian was not a religious man; I am a humanist celebrant.'

Angus looked to where a coffin should have been and saw only an empty space. It saddened him to think that somewhere out in the hills Brian's body lay alone and undiscovered. In place of the coffin was a table with a photograph of a young soldier Angus didn't recognise. A wreath surrounded it, and in front of that sat a beret like those carried by the men at the back of the room, with a row of five medals neatly laid out on the polished wood. It took Angus a moment to realise that the young soldier was Brian.

'He never said he was in the paras,' Angus whispered to Rory.

'He never said much about anything.'

The celebrant spoke of Brian's life, how he'd joined the army and moved north to spend time in the mountains he'd loved. It was a good summary and Angus realised how little he'd known the man.

'Now I'd like to welcome David, Brian's son,' the celebrant said.

David spoke slowly and sometimes struggled to get the words out. 'I remember my dad from when I was a kid. He'd throw me up in the air – you know, like fathers do. And we'd chase balls together. I was small, you know, so it's just the odd memory here and there. He was away a lot in the army, but when he was home we'd spend hours together.'

He paused, and Angus could see the memories playing out in the man's mind: a child's laughter, a ball bouncing away across the grass and big hands reaching out in the air. Then David's face grew dark.

'I didn't know anything about the IRA, I didn't know where Belfast was, I'd never heard of the Falls Road. All I knew was that when my dad came home, there were no ball games, and he didn't catch me in the air any more. He was just this big, silent man.' David stopped, his voice choked, and the chapel fell silent. 'I never really knew him. He just stayed quiet. I won't lie – I couldn't say we were close. He started walking in the hills, and we saw less and less of him. Until, well, you know the rest. I can't tell you any more, but I wish I'd known him better. You see, I lost my dad a long time ago.'

It was quiet then. There were no hymns, but they listened to the Rolling Stones track 'Let It Bleed', which David said was his dad's favourite. The men with the berets filed out and stood flanking the doorway as the little group of hillwalkers left the building. There had been no flowers, and somehow that seemed right – Brian wasn't a flowers sort of person. Folk dropped money into a small tin for mountain rescue.

As the congregation shuffled out, Angus offered his hand to David. 'I'm president of the mountaineering club. I went to a lot of bothies with yer dad. I'm sorry. I wis there when … ' Angus couldn't make himself say the words.

David took his hand and shook it warmly. 'I heard you helped look for him. Thanks.'

Angus could see the empty sleeping bag rolled up against the bothy wall and remembered the horror of finding it empty. He felt a strong urge to tell David. 'I never realised he wis nae back, yer ken. I should have—'

David cut him off mid-sentence. 'He was a law unto himself, my dad. You never knew where he was. Thanks for what you did.'

Angus wondered how much Brian's son knew of what had happened. He wanted to confess, to tell him how badly he had failed his father. Suddenly Jen was beside him, her blond hair bright against her dark coat.

'Nah then. Come away wi' me now.'

'But I need to … '

Jen was firm with him. 'Tha's done all tha needs to do.' She guided him out into the cold rain and Rory and Laura joined them.

As Angus watched the small group of mourners leaving the crematorium Sinead emerged and Angus turned to Rory. 'I dinnae ken she were here.'

'Didn't you see her? She was at the back,' Rory said.

Sinead stood talking to David for a few minutes and then

put her arms around him and hugged him.

The little group of mourners slowly dispersed.

After the service they filed back to the Thistle Inn near the river, where the club met, to drink tea and eat sandwiches and talk about the man most of them had barely known. The club had been subdued after he had gone missing. After mountain rescue stepped down their search, a few of the club members had looked for him themselves, but it had been a fruitless task – they all knew that sometimes walkers vanished in these remote places and were never seen again. A body is a small thing on the vast scale of the mountains.

Angus nibbled at the ham sandwiches and stood in a small huddle with Rory and Sinead, who looked oddly pale in her black dress with her hair tied up.

Angus took a sip of tea. 'Thanks for coming, Sinead.'

She smiled. 'The poor man was lost on our estate, it's the least I could do.'

'It feels strange, having a funeral without a body,' Rory said.

Angus put down the sandwich he was trying to eat. 'Aye, I ken what yer mean but it's only richt that we do something. I dinnae ken he was in the paras.'

Sinead shook her head. 'Jasus, if ma da knew I'd been to a para's funeral he'd go feckin' spare, so he would.'

Angus and Rory looked at her in surprise.

Sinead turned, eyes sparkling with anger, but she kept her voice low. 'Are youse thick? I know where the Falls Road is, an' so did my da.' She glanced over at David, who was talking to Jen and Laura. 'Still, it's no fault of his, is it?'

'No, I suppose it isn't,' Rory said. 'Brian never spoke about it.

Strange how little you can really know someone.'

Angus regretted not having made more effort to get to know the quiet man, but it was too late now. 'He dinnae talk much at all, Brian. Sometimes we'd forget he wis there. Maybe we should have made mare effort.'

Rory shook his head. 'He was a private person. I think that's how he liked it.'

Angus turned to Sinead. 'He came looking for us once.'

'That's right,' Rory said. 'One night when we couldn't find the bothy. It was a foul night too.'

Sinead put down her drink. 'When was this?'

Angus felt more ashamed when he thought of what Brian had done for them. 'Aye, it was one night last winter, yer ken. He came oot with a light and found us. Terrible dreich, it wis. Mare than we did for him.'

Rory turned and put his hand on Angus's shoulder. 'You've got to stop beating yourself up about that. It wouldn't have made any difference.'

'What are youse talkin' about?' Sinead asked with a deepening frown.

Rory answered before Angus could speak. 'On the night he went missing we didn't realise Brian was not back from the hill until the morning. Angus thinks it was his fault but it wasn't. We all should have realised.'

'But naebody did.' Angus bowed his head. 'I'm president, I should have taken more care.'

Perhaps realising that he would never change Angus's mind, Rory turned to Sinead and changed the subject. 'It's the vote on your application for the lynx this week, isn't it?'

'Aye, so it is,' Sinead said with a sigh. 'But we've no chance of getting it through.'

'How's that?' Rory asked.

Sinead brushed back her dark hair. 'Youse told me yourselves that Councillor McCormack is the chair of the committee and Purdey has the fella in his pocket.'

Angus nodded. 'Aye, you're richt – they never vote against him.'

'We never get anywhere. Nothing changes.' Rory spat the words out.

Sinead took a deep breath. 'I can't understand why Purdey is so against the lynx. It wouldn't predate his grouse, and the fella knows it.'

'Are yer sure aboot that?' Angus asked.

Sinead nodded. 'Lynx are forest-dwelling creatures. They wouldn't come out on to the open moorland. They never stray far from the treeline.'

Rory let his plate drop on to the pub table with a clatter. 'He's against it because the blood-sports fraternity oppose any kind of environmental progress. Any victory by us would be a loss to them. They don't want any restrictions on what they do in case it's the thin end of the wedge. So they oppose everything.'

Sinead picked up a slice of cake. 'Lynx live on roe deer, so they do, maybe the odd red deer calf.'

'I bet once they were released you'd never see them again, barely know they were there,' Rory declared.

Angus slipped outside the pub and lit his pipe. As he smoked he began to see a way he might be able to get the application through the council.

Angus left early for work that Monday. Laura was still asleep when he left the house. Normally he would walk the half mile to the sandy coloured, shoebox-shaped offices, but today he took the Land Rover. He wanted to be there early to be in the right position. Angus sat in the driver's seat and pretended to read his newspaper while the car park slowly filled up. He watched as they arrived; accountants and IT technicians, clerks and directors of services. He knew almost all of the 200-or-so staff who worked at the regional headquarters. One by one they all filed across the car park, swiped their identity cards, and were swallowed by the swing doors.

At about nine-thirty a black Mercedes four-wheel drive swung in to the car park, and Angus stepped out on to the tarmac. He realised he was breathing heavily, sweating – even trembling a little. Then he watched himself, as though he were watching a film of someone else, run across the car park and approach McCormack's car. Inside he could see the councillor collecting his papers ready to head in to the committee.

Angus rapped on the passenger window and McCormack looked up. It was only there for a split second, but Angus saw it: an unmistakeable flash of panic when he recognised Angus. McCormack had been attending council meetings for almost twenty years, and Angus had been there all that time. But not once had he met him in the car park.

McCormack froze. Angus tapped on the window again, but this time he smiled, and he watched a ripple of relief pass across the councillor's face as the window slid down.

'Wondered if I might have a quick word, Mr McCormack?'

McCormack beamed, obviously confident that he was on safe ground. 'Of course. Let's go inside and get a coffee,' he said, picking up his papers.

'Nae, here would be much better, yer ken.'

Angus was never the sort of person to menace anyone. All his life, his line had been one of moderation; reasoned debate, compromise and considering the other person's point of view had always been his modus operandi. Considering this it was surprising that, with so little practice, Angus put such menace into those few words that Councillor McCormack began sweating profusely and sank back into his seat without further complaint.

Angus climbed into the passenger seat and smiled again. 'I mind it's the Environment Committee, is it nae?'

McCormack nodded, but he didn't smile; he looked like he was sharing the car with a cobra.

'I was wondering how yer was intending to vote on the lynx rewilding application?'

McCormack said nothing for a moment. Angus could see the gears whirring in the man's mind, no doubt trying to identify a trap.

After a moment, the councillor said, 'Against. Ridiculous idea, isn't it? I mean, what next? Grizzly bears in the park?' He watched Angus's face and suddenly seemed less sure of himself. 'It's a bad idea. Isn't it?'

'Nae, I think yer should support it,' Angus said quietly.

McCormack fiddled with his papers. 'Did that space-age hippy Tony Muir put you up to this? I know you are pretty thick with him after that ceilidh.'

Clearly McCormack didn't think him capable of independent thought. 'Nae, all my idea,' he said, trying not to let his irritation show.

'Well, I'm surprised at you. A council official trying to influence an elected member, very surprised,' McCormack began to bluster. 'Disappointed too, Angus, very. Someone of your seniority breaching protocol like this.'

'Disappointed.' Angus took the word and swilled it around in his mouth as if tasting a fine wine. 'Disappointed. Aye, so wis I when I found oot.'

'Found out what?'

'See if yer can guess, Gerald. Wrack yer brains.' Angus ran his finger over the finely upholstered leather of the car's interior. 'See if anything comes to mind, yer ken.'

McCormack fell silent. 'What is it you are trying to say, Angus?'

'I always wondered how yer did it. All those weekends when you slipped away from your wife to sleep with mine.'

The colour drained from McCormack's face, as if he had just felt the fangs of the cobra he'd been sharing the car with sink its teeth into his skin. 'Ah, I see.' Clearly there was no point in denial. 'How did you find out?'

Angus stared ahead, speaking as though he were explaining how he had learned to grow roses. 'It were little things. Twa wine glasses in the dishwasher when there should only have been one. The way Laura was always keen for me to bide awa' at weekends. But I dinnae really ken till I saw yer on the lawn at yon Muir's hooley. She reached oot and touched yer cheek and I mind she used to touch me like that.'

McCormack stared out of the car window for a long time before he spoke again. 'I suppose you'll be telling all and sundry now.'

'At first I wis so angry I was afraid I might burst, but then I got to thinking. If it makes her happy then maybe's there nae so much harm in it. An' I wis thinking we might come to an arrangement.'

McCormack tuned his eyes wide. 'What do you mean – blackmail?'

Angus shrugged, a little offended by the phrase. 'I widnae call it that. But I suppose there's them that might. Yer get the Muir estate's application for lynx rewilding through, and I'll no clype on yer, ken.'

'That's it? That's all you want? Just a planning consent?' McCormack looked relieved that he was not being asked for more.

This provoked Angus's anger for the first time in their little chat. 'That's all it is to yer, isn't it? Jis' another planning application that yer pal Purdey wants stopped. Well, it means something to me. Do yer want me to ask for more?'

'No, no, it's fine. I'll vote for the application,' McCormack said hurriedly.

Angus shook his head. 'Nae, Councillor, I didnae ask yer to vote for it, I asked yer to get it through.'

'But all I can do is vote,' McCormack protested.

'Yer can do a *lot* more than that. Yer ken how things work. Call in a few favours, twist a few arms, bribe someone. Just get it through.' Angus stepped out of the car and left McCormack staring at him. 'Just get it through, yer ken.'

Lord Purdey strolled into the council chamber and took his seat in the public gallery. He placed his bundle of papers carefully on the polished wooden table in front of him and smoothed out the folds of his tweed suit. That morning he had dressed with care. He had toyed with the idea of wearing a business suit; after all, this was a business meeting of sorts and he would be a long way from the open moorland of his country estate, so there was no need to wear anything capable of withstanding the vagaries of the Highland weather. Eventually, however, he had decided that those attending would expect to meet with the owner of a country estate, and he had dressed in a new tweed suit he had been saving for a special occasion. The weight of the tweed and its familiar rough texture made him feel comfortable and in charge. At his wrists were the cufflinks his grandfather had owned: nine-carat gold with the Purdey emblem, a peasant being beaten by a lord, engraved upon them.

Purdey cast his eye over the council chamber, which was now beginning to fill up with the elected members taking their seats. The chamber was arranged rather like an amphitheatre, with the councillors seated in a semicircle above a sunken area with a lectern, and behind that were desks for the chair of the committee. McCormack had yet to arrive, so his chair stood empty. Purdey noticed a bearded middle-aged man taking his seat next to where McCormack would sit. He looked like the sort of chap who would be taking minutes. Purdey stared at the man. There was something familiar about him, and the more he looked at that grey bearded face the more he felt that he had seen him somewhere before. But where?

Over on the other side of the room Purdey sighted the enemy: the young woman employed by that hippy Muir as his estate manager. Muir had made a mistake there. Clearly it was a man's role; none of his estate workers would listen to a woman. Worse than that, she was Irish. Muir was new money, and his ignorance was showing. Purdey fiddled with his cufflinks and sighed. *What a shame this rigmarole has to be gone through at all.* The application would be rejected – everyone in the room knew that. Even that Sinead woman couldn't be totally stupid. She must know she was wasting her time. They would listen to her application, smile, listen to the objections, talk for a few minutes and then vote against it. The whole thing was already decided.

As Purdey looked up Gerald McCormack entered and walked swiftly to his seat at the top table. Purdey smiled at him but the councillor averted his eyes, stooping his tall, thin body awkwardly to sort his papers on the desk. Purdey realised instantly that the councillor was playing a very clever game. It would be foolish to acknowledge one of the objectors at this stage in the proceedings as it might give an impression of bias. McCormack had never advertised that he was the Purdey estate's solicitor, yet he had not hidden it either – sensible fellow. Bias was everywhere anyway, but one didn't draw attention to it.

Confident, Purdey beamed across the floor of the chamber at the Irish woman. She made eye contact but didn't return his smile. Purdey relaxed, satisfied that he already had the day won, but by the time they began the proceedings he was already bored and looking forward to his lunch date with McCormack. In truth he didn't much like the man but there was no denying that he had his uses.

Sinead was invited to speak in support of her application and soon her Belfast accent filled the council chamber. While she spoke, Purdey took out his phone and read his texts. There was a message from Tabatha reminding him to pick up her clothing order from one of the shops in town and a few good-luck messages from other members of the campaign. He glanced up. Sinead was showing a graph on the big TV screen at the front of the chamber; then she showed maps of where lynx roamed free in Europe. Purdey paid only scant attention. The poor girl was wasting her time and he was bored with the whole proceedings.

Sinead put a picture of a lynx on the big screen. 'Lynx continue to live in the wild in areas of Spain which are much more densely populated than the Highlands. There are very few incidents of predation on domestic animals and no instances of attacks on people.'

Purdey glanced at his watch. Twenty minutes had passed since he last looked at it. On the screen above her a lynx stared out into the council chamber, its yellow eyes glowing in the dimly lit chamber. *Blasted creature*, Purdey thought.

Sinead was summing up, he realised. 'So I urge you to grant this application and allow this magnificent animal to return to its home and make the Highlands a richer place.'

'Codswallop,' Purdey murmured under his breath.

'Objections?' the bearded official announced, and rummaged through his papers. Then he read out: 'Charles Purdey, Campaign Against the Persecution of Landowners.'

Purdey rose and strode to the lectern and as he did so he found himself staring at the face of the council official. An image flashed into his mind of three naked men standing on the riverbank of his estate. He looked again at the official. *Could this chap have been one of them?* The thought so startled Purdey that he dropped his notes. As he bent to collect them he dismissed the idea, and he glanced around at the expectant faces of the councillors, most of whom he knew personally. People whose votes he could rely on. He kept his best smile for Councillor McCormack, but even now his smile was not returned – and more than that, this time he saw a look of doubt in the councillor's eyes, perhaps even fear.

Purdey dismissed the thought, placed his thumbs in the pockets of his yellow waistcoat, and began.

'Now look here, as you all know it's been 1,300 years since these animals roamed free in Scotland. We simply can't turn the clock back and expect there to be no problems. What happens in Spain is all very well, but this is not Spain. Just because no lynx has ever attacked people doesn't mean it's impossible. How do we know it never happened in the past? Thirteen hundred years is a long time – for all we know there might have been lynx dragging away children all those years ago on a daily basis. We can't be sure, can we? And if the Highland Council votes to reintroduce these vermin and a child goes missing, well, fingers would be pointed. And we all know where.'

He went on to paint a picture of the destruction of sheep farming, predation on a mass scale. All the time he spoke he looked straight at Councillor McCormack, but the man would not meet his gaze.

After his speech Purdey returned to his seat, patting a few of the elected members on the shoulder as he did so. A few other letters of objection were read out. One was from a group of farmers fearing for their livestock and another was from an elderly lady objecting to the construction of a golf links.

The bearded official turned to McCormack. 'Shall we put the application tae a vote, Mr Chairman?'

The colour had drained from McCormack's face. 'Yes, quite so,' he said with the air of a condemned man.

As the vote was announced McCormack made eye contact with a number of councillors and raised his hand for those in favour. A significant number of other councillors put up their hands.

'Against,' the official called, and counted the hands that went up. 'The application is approved.'

There was a stunned silence in the room. Purdey stared across the chamber at McCormack and made a passable impersonation of a goldfish. He felt his breath coming in short bursts. *What the devil? Surely they won't listen to that Irish girl.* He couldn't believe what had just happened. He wanted to yell out that there had been some mistake.

At last he hauled himself to his feet, legs trembling, face hot. As he hurried from the chamber he paused only briefly to fire a poisoned look at McCormack.

'You've not heard the last of this.'

***

Angus was walking across the car park when he saw Sinead running towards him, a broad smile across her face. 'Jasus, we feckin' won! I never expected that now. I was certain we'd lose.'

Angus grinned back. 'Aye well. that's democracy for yer.'

'All we need do now is get it rubber-stamped by the Scottish

Parliament and we can bring back the lynx.' Sinead pulled her
phone out of her bag.

'I cannae help yer there.'

Sinead's triumphant smile wavered for a moment. 'What do
youse mean, help us?'

Angus realised he had said more than he intended. 'Jis' that
I'm sure yer'll need nae help with them.'

Sinead laughed. 'Aye, I'll phone the wildlife park. They'll
be delighted – they have a pair they're keeping away from the
public so they don't get used to being around people. Should be
ready to be released soon. It's the ideal time of year for it, so the
experts tell me.'

***

Autumn arrives quickly in the hills of the Highlands. September
whispers of changes to come; then October arrives, like a terrible
child, running through the glens spraying the forests with gold
and staining the hillsides yellow and brown. But the glory of
the forests in autumn is brief; October brings with her gales
that tear at the limbs of the trees, leaving them veined black
against the grey skies. Then the early frosts of November make
all nature brace itself for the godless season, for soon winter
will stride across the landscape, bringing the cold silence and
wrapping the dead hills in a shroud of white.

Rory finished the lentil bake and sat cross-legged on the
settee. Jen was sleepy. Her long shift at the hospital over, she
lay with her head on Rory's lap and he played with her hair
while she dozed. He flicked on the TV just as the Scottish news
was starting. A reporter identified as Simon Partington was

standing in front of the Scottish Parliament building, the wind catching his long dark hair. When he spoke, his voice almost woke Jen and she groaned a little, so Rory turned off the sound and switched on subtitles.

'Today the Scottish Parliament is voting on the Highland Council's plans to permit the reintroduction of lynx,' Simon said. 'The plans are controversial and many landowners are against the move, including Lord Purdey of the Campaign Against the Persecution of Landowners.'

Lord Purdey appeared on the screen, straightening his grey moustache. Behind him stood a ten-foot-high model of a lynx with a dead lamb in its mouth. Blood was dripping from the huge fangs tearing into the lamb while menacing eyes stared into the camera. Rory thought it looked more like something out of *Doctor Who* than a wild animal.

Simon thrust the microphone into Purdey's face.

'Lynx have been extinct here for 1,300 years,' Purdey said in a condescending tone. 'Many who actually live in the countryside fear that if it returns it will take a terrible toll on our livestock, and, who knows, even take the odd child.'

'But that's never been known to happen, has it?' the reporter interjected.

'But that doesn't prove it *won't* happen. Ha! Look at those teeth.' Purdey pointed up at the huge fangs of the fake animal.

'Some supporters of the proposal were also here today,' Simon continued. 'Sinead Callaghan, why do you think reintroducing the lynx to the Highlands would be a good thing?'

Sinead looked a little nervous. 'Well, these animals should be part of our natural flora and fauna. They belong here. I think youse should be asking why *not* reintroduce the lynx, so you should. They would also control the fox population and keep down roe deer.'

Purdey's voice could be heard in the background. 'Bloody sandal-wearing vegans.'

'But concerns have been raised about the threat to people and livestock,' Simon pressed.

Sinead looked irritated. 'All that is covered in the proposal. Lynx avoid human contact. We might lose a handful o' sheep but there'd be compensation, so there would.'

The TV cut to discussions in the Scottish Parliament. Nicola Sturgeon appeared on the screen. 'We've listened to the concerns of local people, and the Scottish Government has no plans to reintroduce the lynx.'

Simon looked out of the screen. 'The proposal has been voted down in the Scottish Parliament.'

Rory picked up the remote, switched off the TV, and hurled the control across the room. 'Oh shit, shit, shit, shit!'

Jen jolted awake and looked up at him with sleep-filled eyes. 'Wa's up wi' thee?'

'They've turned down the lynx rewilding proposal. Ignorant bastards.'

Jen lay back down across his knees. 'Oh bugger.'

Rory reached for his phone to call Angus but it rang just as his fingers touched it.

It was Angus. 'Rory, did yer see the news?' He sounded angry.

'Yes, Angus, I saw it.'

There was a pause, then Angus spoke again. 'Meet me in the

pub, around nine.'

Rory was surprised – it was Tuesday and they only ever met on Thursdays. 'Oh, OK. I'll bring Jen.'

'Nae, jis' you. Leave Jen at home, yer ken.'

The phone went dead.

***

Rory pushed open the door of the lounge bar of the Thistle Inn. The little lounge was almost empty, with only two couples sitting chatting over drinks. Angus was not at his usual table in the centre of the room. Instead, Rory found him in the far corner, away from the other drinkers.

Rory placed his glass on the table and sat opposite Angus. 'I can't believe it. We got it through the council only for the bloody parliament to turn it down.'

Angus laughed. 'Except we didnae.'

'What?'

'The vote was fourteen against the application and thirteen fer.' Angus smiled as he sipped his beer.

Rory stared at his friend. Angus was always so proper, always played by the rules – he just couldn't believe he'd have cheated. 'You mean you fixed it?'

'Nae, I must have miscounted, yer ken. Easy mistake to make.'

Angus beamed with pleasure, but Rory barely knew what to say.

'Bloody hell. How did you get away with that?'

'Easy,' Angus said, eyes twinkling. 'I've been doing it for twenty years, counting the votes. Nae one ever checks. I'm always right, yer ken. In twenty years I've never made a mistake.'

'You have now.'

Angus raised a finger, as if making a point of order. 'Nae, I dinnae make a mistake, and I mind I got the vote richt.'

Rory tugged at his scraggly beard, the way he always did when he was struggling with something. Angus looked different somehow – determined, confident – as if he'd finally found something to fight for after the hopelessness following Brian's disappearance.

'They always win, though,' Rory said. 'When it matters they win. They got to the SNP over planning permission for hill tracks, didn't they? They approved those hydro schemes in Glen Etive. All seven of them. Including those in the Wild Land Area. It's a bloody joke.'

Angus seemed subdued again. 'Now the lynx. We'll nivver beat them, not legally.'

'We'll go back. We'll try again and eventually we will win.'

'Nae. It's their system. These dukes and lords and landowners. They own most of Scotland atween 'em, and us peasants get nae say.'

Rory took a sip of beer. 'I don't think I'll ever see wild lynx in Scotland, not in my lifetime.'

Angus snorted. 'Aye, they block everything we try and do. It their land, yer ken. They can bulldoze tracks all over it, blast wildlife to hell, burn the bloody heather. And there's nae one thing we can do aboot it.'

'We just keep on trying. We'll get there. We got access to the land, they tried to stop that.'

Angus shook his head, growing eloquent with anger.

'Do you know how we got access rights?'

'Not exactly.'

'We got them because folk wis prepared to break the law, to go to prison so they could walk on the hills. I read aboot it. In 1932 the landowners were trying to stop people walking on the hills an' a bloke called Benny Rothman organised a trespass doon in England, a place called Kinder Scout. They marched and got arrested and some of them went to prison. Only then did the landowners cave in, and that's because they could see they were going to lose.'

But Rory shook his head. 'We'll never see lynx free in Scotland.'

The pair lapsed into silence, frustration crackling in the air around them.

Then Angus spoke. 'Yer richt. It'll nae happen in my lifetime or yours. There's nae way they'll let it happen.'

Rory nodded. 'We won't live to see it.'

Angus was suddenly grave, as though he had come to some serious conclusion. 'Not unless we do something aboot it.'

CHAPTER 12

Rory and Angus sat in the Land Rover and watched as the white deck of the *Maid of Morvan* inched its way towards them through the mirror-calm water. It was still early, only a few minutes past seven on this very cold November morning. Jen would still be at work, Rory knew. The sun had yet to rise, and so the ferry's navigation lights danced on the water as it approached through the pre-dawn twilight.

Angus turned to Rory. 'Once we're on board yer better keep that ferryman talking. I dinnae want him wondering what's underneath yon tarpaulin, yer ken.'

Rory glanced towards the metal cage on the trailer behind the Land Rover. 'It's tied tight enough. I've checked.'

The Land Rover came to a halt on the ferry and Angus switched off the engine. They were the only passengers that morning. Rory got out of the vehicle, trying to look as casual as he could, and zipped up his fleece against the cold air. He strolled over to the ferryman, who was smothered in orange waterproofs over layers of threadbare pullovers, prepared for the long, cold day ahead.

Rory smiled. 'Morning.'

The ferryman scowled at him with bloodshot eyes. 'Jesus, you two are early, are ye no? What time did you leave Inverness?'

'We were up at four to get an early start for our walk.' He

passed a small pink ticket to the wet hands of the ferryman.

The ferryman scratched beneath his woollen hat. 'I dinnae think I were in bed much before that.'

Rory laughed. 'Big night?'

The ferryman stretched lazily as the boat began to head back to Morvan. 'Och, jis' a Saturday night in the old inn, same as usual.' He glanced over at the Land Rover. 'It'll be an extra four pounds.'

Rory was puzzled.

The ferryman pointed at the cage. 'For the trailer.'

It hadn't occurred to Rory that they would have to pay extra over and above the book of pink tickets that they used every week to cross to the island. 'Oh yes, give us a minute.'

He ran over to Angus, not wanting the ferryman to get too close to the trailer and its contents.

'We need another four quid for the bloody trailer.' Rory glanced over his shoulder and the ferryman was already making his way over to the vehicle.

'I've nae cash.' Angus was searching frantically in the glove compartment.

'Shit, me neither. He's coming over,' Rory hissed.

Angus fumbled with a handful of change he'd found in the corner of the glovebox. 'Two pounds seventy-six.'

Rory searched the pockets of his hill trousers. He went from one pocket to the next, finding nothing. For some reason the makers of outdoor equipment think it their mission in life to put as many pockets as possible in hill trousers. This seems particularly odd as most hillwalkers carry a rucksack to put their possessions in, so they actually need *fewer* pockets than most people. These multiple pockets are a curse on the outdoor public, as Rory was discovering; it was only when he explored the seventh pocket, cleverly hidden in his pants, that he located a handful of change.

Rory passed the jumble of loose change to the ferryman, who sighed mournfully as he slowly counted out the assortment of coins.

'Cold today,' Rory announced in an attempt to keep the man engaged.

The ferryman looked at Rory sadly, but didn't say a word. Rory realised that a person who lives his whole life partially immersed in water must have long ceased to notice the weather; maybe this kind of small talk wasn't his forte. The ferryman handed Rory back a two-pence piece that he had obviously decided was surplus to requirements and then glanced curiously at the trailer.

'What you got in there?'

Rory frantically searched for an explanation. 'Pigs, we got two pigs in there.'

'Looks a bit dodgy to me,' the ferryman grunted.

Rory panicked. 'Does it?'

'Aye, that tyre's a bit flat,' he said, and he kicked the tyre of the trailer. There was a low growl from inside the cage. 'Those pigs just growled! I never heard of pigs growling.'

'Yes.' Rory's mind was at full speed now. 'Yes, these do. They're Welsh growling pigs.'

Inside the Land Rover, Angus rolled his eyes.

'Welsh growling pigs? I never heard of 'em. Can I take a look?'

The ferryman began to lift the tarpaulin.

Rory grabbed the sheeting and pulled it back down. 'Sorry, no. You can't. You see, the growling pigs are nocturnal. They get upset in the daylight.'

The bow ramp descended and hit the concrete slipway of the Isle of Morvan with a crash. Rory jumped back in, and Angus put the vehicle into gear. They drove off the ferry and on to the island.

Rory sighed with relief. 'God, that was close.'

'Welsh growling pigs! Nocturnal! That the best yer could think of?'

'He bought it, didn't he? I couldn't think of anything else.'

The roads were quiet that Saturday morning and as they drove the dozen or so miles to the Muir estate they were grateful for a mist that rolled in from the sea and shrouded the forests on the island. At last they reached a locked gate across a track that led into the forest. Rory stepped out on to the gravel road and unlocked the padlock with the key he had obtained from a forester (thanks to a bottle of the Ardmore) a couple of weeks ago.

Soon they were driving up the hill and into the forest. At last they could relax, hidden from any prying eyes. Angus brought the Land Rover and its trailer to a halt in a small clearing where the oak trees had been blown down in an October gale.

'What do you think?' Angus asked.

Rory looked about him and sniffed, the smell of damp trees filling his nostrils. 'As good as anywhere, I suppose.'

They both got out of the vehicle and walked quietly to the back of the trailer. Rory realised that he was trembling. Angus lifted up the tarpaulin covering the rear door, and both men kept to the side, out of sight of the creatures in the cage. He pulled the bolt out of his side of the door. Rory did the same.

Rory looked at Angus over the top of the cage. 'Right then.'

Angus swallowed and nodded. 'Richt.'

Together they lifted up the door of the cage and stepped back towards the Land Rover. There was a hiss from inside, and something stirred, scraping against the metal, but nothing appeared for a moment. Then they heard a menacing snarl and watched as a large spotted creature padded slowly down the ramp. Powerful muscles tensed beneath its fur. It paused on the ramp and turned to look at Rory; he could feel its eyes boring into him, black pupils in wells of gold. The big cat growled and Rory thought it was about to attack. Instead it sniffed the air and walked casually a few feet away from the cage.

Rory glanced over at Angus, who looked even more tense than he felt. Rory whispered, 'Maybe we should get back into the Land Rover.'

Angus shook his head and Rory noticed that he was hanging on to the mesh of the cage with all his strength.

The big lynx glanced back at the two men for moment and then turned and headed for the trees, moving with the gentle grace of a creature born for the forest.

There was a moment of silence, then a hiss, and the second cat leapt from the cage, landing softly on the gravel of the drive. For a second it froze, taking in its surroundings. Then it too bounded for the trees, its grey-spotted fur rippling as its body shifted over the contours of the ground. Soon they were both at the edge of the forest. One of the beasts turned and for a moment its eyes shone yellow fire against the darkness of the trees. Then they slipped away, becoming shadows in the twilight of the forest, and just like the mist in the morning sun, they were gone.

Rory and Angus watched as the cats disappeared. As they left a sudden breeze caught the tops of the trees and they sighed in the wind. *Ah, my children, where have you been?*

Rory turned to Angus, his eyes alight with joy. 'We did it, we bloody did it.'

'Aye, damn it, we did, didn't we?'

Rory could feel tears on his cheek. 'There are lynx roaming free in the hills of Scotland again.'

Then they danced and hugged each other.

Only the lynx knew what had taken place, and they are the keepers of secrets.

✳✳✳

By the time they entered the clearing and Glen bothy came into sight Angus was beginning to feel weary. He had been up most of the previous night and the tension of stealing the lynx had drained him. He thought that Rory looked tired as well. When they opened the rough bothy door they found eight university students; they looked as if they had spent the previous night sitting up late drinking and smoking, and were still milling about, half asleep and hungover. They reminded Angus of how he had been on his early trips before he learned the benefits of moderation.

Rory cleared away some of the cans and bottles to make a space on the table. 'I could murder a brew.'

Minutes later the kettle was steaming and they took their tea outside to sit on the stone wall where they could talk in private. Angus took out his tin of Irish Cask and slowly filled his pipe. There was an unusual stillness in the glen that day. No birds were flying past, and Angus could only hear a few calling. The house martins he and Rory had watched the night Brian had vanished would be in Africa by now. Only the conical nests made of mud, wedged beneath the eaves of the bothy, remained as evidence of these summer visitors.

Angus lit his pipe and let the rich aroma of tobacco fill his nostrils. 'We did it, then.'

Rory sipped his tea and nodded. 'I can't believe it was your idea.'

'I've been working for the council for mare than thirty years, an' I got to wondering, ken – what could I say I'd done in all them years?'

Rory nodded again, and then his expression became mischievous. 'We're criminals now. Wanted men.'

Angus blew smoke into the air and cried, 'Outlaws!'

The idea tickled both of them and they shared a laugh.

'Seriously though, what happens if we get caught?' Rory said, suddenly grave. 'Theft of a couple of lynx. Got to be serious, hasn't it?'

Angus shrugged. 'Maybe they'll nae catch us.'

Rory let his empty mug fall into the grass. 'You'd lose your job. Perhaps we'd get jail.'

Angus sighed and busied himself tending to his pipe. 'Probably. But yer ken we wis talking aboot Benny Rothman and yon Kinder trespass?'

Rory nodded.

'Well, I got to thinking – it were only after he broke the law that things changed. He went to prison – why not me?'

Angus tried to sound confident when he talked about what

they had done, but secretly he was terrified. He could be arrested and charged and the career he'd put together so carefully over the years would be in ruins. A dark shape loomed overhead and Angus looked up as two ravens flew past, their black bodies silhouetted against the grey clouds. Soon they would be the only birds in the glen, making a living by their quick wits and ability to adapt to any opportunity. They reminded Angus of the black robes of a judge. Part of him shivered a little.

Rory pulled out an OS map of the area. 'What was it Donald said? Follow the river.' He stared at the map for a few minutes. 'There's nothing marked as far as I can see. Take a look.' Rory passed Angus the map.

Angus let his eyes follow the river as it wound its way across the open hillside of the Purdey estate and then on past Glen bothy until it met the forest, where it went on for four or five miles before finally meeting the sea at a small bay. He searched for any sign of a building, even the mark of a ruin, but there was nothing.

At last he folded away the map. 'Nae, I cannae see anything. Unless we can get them deeds Purdey'll take this bothy back for sure.'

Rory got to his feet. 'Better see if we can find the place, then.'

By the time they reached the riverside the mist of morning had cleared and given way to a brisk cold wind. The oak trees still clung to a few of their autumn leaves, keeping the forest alive with colour, but soon November would come to an end and December would strip the branches bare. At first they found their way easily, weaving between the trees, their boots hardly making a sound on the soft mossy earth. Soon, however, the trees closed in around them, shutting out the wind and filling the air with the smell of wet moss and lichen. Now they were scrambling over fallen boughs, tripping over half-buried roots, moving slower and slower as the undergrowth closed in.

Rory looked about him at the tangle of trees. 'I wonder when the last person was here.'

Angus nodded. 'Aye, maybe a hunter in the last ice age judging by all this growth.'

They walked on for a few minutes in silence. 'I wonder where those lynx are.'

'Separated by now, maybe.' Rory paused to wipe the sweat from his face. 'Lynx are solitary, they only come together to mate.'

Angus looked at the tangle of trees and bushes surrounding them. 'How the hell do they find one another in all this guddle of forest?'

'Scent trails, I expect. I'm pretty sure they don't call to each other.'

They pushed on into the forest. There were no tracks here, no paths. Sometimes they picked up a faint deer track weaving its way through the woods, only to lose it again yards later. After a while of hacking through this thicket, Angus decided it was time for a rest.

He sat down on a fallen tree. 'We could pass within twenty feet of an old bothy and nae see it in this jungle.'

Rory sat down beside him and began searching in his bag for some fruit, but Angus rummaged in his own rucksack, produced some chocolate raisins, and handed the bag to Rory.

Rory looked at the bag. 'Whey powder and milk. I better not.'

Angus chuckled. 'Go on, min. Yer a lynx rustler now. Have a chocolate raisin. Wild rustlers don't fret about the odd chocolate raisin.'

Rory laughed and popped a handful of raisins into his mouth. He savoured the taste. 'Oh wow. It's been so long. Milk chocolate! Forgive me, Father, I have sinned.' Rory chewed the raisins thoughtfully. 'Thirty years, that's a long time.'

Angus nodded slowly. 'Six months ago I could nivver see myself doing anything like this.'

Rory took another few raisins. 'Well, you've done something now. You're the man who released the lynx.'

'One of 'em.'

'It were your idea.' Rory chewed the raisins enthusiastically.

'Aye, but yer didnae need much persuading. What do yer think it's like, being locked up?'

Rory swallowed his raisins. 'No trees, no hills, no wind on your face. Just bricks and bars and walls. I don't know if I could stand it.'

Angus got to his feet, trying to push back the fear and hoping Rory wouldn't notice. 'Aye well, let's just keep our heads down. You ready?'

'Come on, then. Where's this bloody bothy?'

They walked on for two more hours but the old wood kept its secrets and they found no sign of the building. The forest was so big and a bothy such a small thing that the chances of finding it felt tiny. At length they decided to head back to Glen bothy and give it another shot in the morning. As they made their way slowly back through the forest, following their tracks as best they could, the skies turned gradually darker and a light, cold rain began to fall. When they got back to Glen bothy the students had left, an assortment of empty cans and bottles marking their passage. The old building had returned to silence.

Angus looked at the mess. 'Why cannae people take their stuff oot with them?'

Rory was already gathering the bottles up into a plastic bag. 'You were young once, Angus.'

'Aye, I was ken. An' I nivver left bothies full of bottles.'

The shadows were lengthening, the rain getting stronger. Already it was near dark inside the bothy. Angus lit a few candles and the world shrank to the small area of the room reached by the warm, flickering glow. He lit his pipe and everything outside the bothy's walls receded to a faded memory as he let the smoke slowly drift into the empty air.

Rory opened a tin of beer. The crack sounded loud in that quiet space. 'I heard they found a bothy not so long ago. They felled a forest somewhere and there it was just sitting there, lost for over twenty years.'

'Odd, isn't it?' Angus said, blowing a smoke ring. 'A place might have been someone's home once. Time jis' kinda swallowed it. Folk forget.' Angus blew a smoke ring. 'Are yer any closer to finding an island to live on?'

Rory swallowed a mouthful of beer and scratched at his beard. 'Not really. I've found a few off the west coast that might be good. Perhaps next year, if we can get the cash together. I've got to get out of BetterLife. It's driving me up the wall.'

Angus leaned back, puffing gently at his pipe and letting the smoke sooth him. 'Aye, I ken how that place must get to you. We're nae meant to be machines.'

Angus sensed that there was something Rory was struggling to say, so he sat quietly with his pipe.

At last Rory spoke. 'Thing is, you know, I'm not sure if Jen's really that keen.'

Angus didn't know how to respond. Who was he to give relationship advice? 'Do yer ken if she nae wants to leave the hospital?'

Rory shrugged. 'She told me she's doing OK there, said it might be a long time before we could move. I don't know.'

Angus felt uncomfortable, as if he'd pried into a sensitive area, so he changed the subject. 'Maybe we should light the fire.'

Rory grabbed a box of matches and lit the firelighters they'd carefully layered beneath the coal. They watched as cheerful yellow flames began dancing amongst the dark coals. The fire gave them something to watch as it gradually took hold. After a while, Angus realised that he was beginning to feel hungry. He set out his stove on the table and was getting the fuel out of his rucksack when he heard the rough wooden bothy door rattle as the bolt was drawn back. The door creaked open and a man stepped into the room.

It was Brian.

***

The pair rose together and stood transfixed. Angus stared at the figure in the doorway – it was definitely Brian, but he looked different, more solid and colourful than he had before. He had put on weight and stood taller than Angus remembered.

Rory knocked over his beer can and stammered. 'Bloody hell, Brian. We thought … Where have you … '

Angus decided that it was better to let Brian talk at his pace, and placed his hand on Rory's shoulder to silence him. 'Come in, Brian. It must be thirsty work being dead. Yer look like you need a dram.'

Brian sat cradling a whisky. Angus knew better than to question him. Brian was too private a man for that. Better to wait until he was ready. Angus felt a great wave of relief crashing over him; he didn't say anything but every part of him was cheering, and it felt like a colossal weight had been lifted from his shoulders. No one comes back after three days, and Brian had been gone for a lot more than that. He wanted to tell Brian so many things – how sorry he was that he hadn't noticed he'd gone, how they'd searched for him, how desperate they had been to find him, and how much he regretted that no one had taken the time to really talk to him. Angus knew that second chances are rare in life. Broken relationships, harsh things said, cannot normally be undone; yet here was Brian, sitting beside the fire, smiling at him.

At length Brian spoke. 'I thought you'd be here. I must have caused a lot of trouble. I'm sorry for that.'

Angus put his hand gently on Brian's arm and was surprised to find tears welling in his eyes. 'Dinnae worry. Yer all richt, that's all anybody's worried aboot.'

Rory sat quietly listening, letting the two older men talk.

'It's all right, Brian,' Angus whispered.

Brian took a long breath, like a swimmer about to dive into a deep, dark pool. 'I was on the summit. Everything was still and quiet. I turned to come back here but my feet just wouldn't move. I tried but my feet stuck solid.'

Angus could see that Brian was struggling. 'Brian, yer don't have to tell us.'

'No, Angus, I do, I bloody do – I should have told people years ago.' Brian took a gulp of whisky. 'I walked away. Away from the noise, the things I've seen, the nightmares.'

Brian stared down at the floor. His body shook and he squeezed the cup so hard that Angus thought it would shatter.

Angus spoke softly, afraid his words might cut Brian, but wanting to make whatever he was trying to say easier. 'Saw in Ireland, yer ken? Northern Ireland?'

Brian didn't answer that but his body relaxed a little. 'My legs just took me away and I walked, just kept walking and walking. I've been walking away for years, all these years. I saw the helicopter searching. How could I tell them me feet wouldn't bring me back?'

Brian stared into the fire. His eyes filled with tears, and then he turned to Angus. 'How did you know about Ireland?'

'David told us,' Angus whispered.

'David? He was here?'

Rory broke his silence to explain. 'We had a service. Memorial like, and he came.'

Brian suddenly understood and he laughed. 'You never did. For me?'

'Yes. There were paras there.'

Brian's laughter died. 'Paras? I suppose there would be. David ... I was never much of a dad to him, you know.'

That night he spoke more in a couple of hours than Angus and Rory had ever heard him say in his life. Now the words tumbled out of him.

'After a few days I came to a cottage. I was hungry and there was a woman there, Kathleen her name is. She fed me and she didn't ask me anything, no questions, and I felt … ' He struggled to find the right word. 'I felt at ease with her. Peaceful. The days passed and we got to liking each other. I never thought I'd find another woman again, but Kathleen … she just seemed to know when I needed peace.'

Brian laughed again and Angus realised it had been a long time since he had heard Brian laugh. 'She fed me, you know. And I ate. I haven't eaten much in a long time. Not since the army. Look at me – I've put on weight.'

He grinned and patted his stomach.

'Yer have so, Brian,' Angus said gently.

'It was a stupid thing to do,' Brian said slowly. 'But it turned out to be the right thing. I don't know if that makes sense.'

'It does.' Angus let the silence rest in the air. 'Are yer coming home?'

Brian shook his head slowly. 'No, not to where you mean anyway. I am home, Angus. I have a proper home now with Kathleen. I've been working in her garden, growing food. We don't need much, the two of us. There's more than we need, really. I thought you should know. I don't want to worry people.'

Angus could see a contentment he'd never seen in Brian before. 'Aye, I'll tell them, Brian, I'll tell them yer OK.'

Brian got up to leave.

'I'm so glad I found Kathleen's cottage. I spent the first three nights in an old bothy in the woods.'

Rory sat upright with surprise. 'Do you think you could find it again?'

Brian showed them on the map where the old bothy was, and then went back to his Kathleen.

***

Rory and Angus sat up late, talking and drinking too much whisky. It had been quite a day; they had become lynx rustlers and seen Brian come back to life all in the last twenty-four hours.

It was well past midnight when they decided to head for bed. Angus stood unsteadily as Rory blew out the candles. 'I'm awa' for a pee.'

It was cold outside the bothy. The rain had ceased and the clouds parted to reveal the shining face of the moon. Angus shivered as he stood finishing the last embers of his pipe, watching the clouds scudding across the night sky. He turned to go back to the bothy and James was standing beside him. Angus was surprised at how young he looked with his blond hair blowing in the moonlight.

James thrust his hands into his jacket and looked up. 'It's a fair moon, hey?'

Angus looked up at the shining disc. 'Aye, that it is.'

James turned and gave Angus that big grin of his. 'It wasn't your fault, Angus.'

Angus answered him slowly. 'Are yer sure of that now, James?'

James laughed again. 'Aye, it was nobody's fault.' James shrugged and began to whistle as he walked away toward the woods, but then he turned. 'Go on, get yourself in. You'll catch your death out here.'

CHAPTER 13

The big cats moved cautiously, unsure of their new environment. It was a cold night and stars twinkled between the skeletal fingers of the trees. Samson, the male, led the way through the forest, his breath misting in the cold night air. Here and there shafts of moonlight cut the deep darkness. The night was hung with silence. To human eyes the darkness would have been impenetrable, but for Samson the forests gleamed with detail, and scents told him stories of the creatures that had passed that way. He searched the trunks of the trees, the thick undergrowth and the fallen limbs for signs of movement. He was about the height of a large Labrador – larger than the female, Delilah – and his rear limbs were heavily muscled so that they could propel him forward with immense speed to ambush his prey.

Delilah followed. She was lithe and agile, and her eyes burned with a fierce intelligence. She sniffed the night air, catching the heavy reek of a fox who had passed by only hours before. Then she sensed the signature of another animal. It smelt of grass and wet fur. She stopped as Samson bounded over a log, then she pulled the air more deeply into her nose. The scent triggered a deep instinct in her. This animal she could smell on the wind was prey.

She watched Samson as he padded silently and methodically through the forest. Each time they moved forward they expected to find hard steel mesh but all they found was forest. They had been in captivity most of their lives, had grown used to there being barriers around them, limits on their wanderings, but now they probed for the edge of their cage and found nothing but freedom.

In the wild they would have led solitary lives, coming together only to mate, but the cages they had lived in had made them accustomed to each other. As Samson passed a tree there was a sudden explosion of movement above him and he spun, hissing, as a startled barn owl took to the air. The lynx watched as the owl drifted away in a graceful arc.

They moved like ghosts through the trees, passing over fallen trunks in a secret silence. For hours they padded through the forest, crossing streams and once or twice breaking out on to the open hillside. They were driven on by hunger – instinct told them that they must hunt or die.

At last they came to the edge of the forest where the moorland stretched away in the moonlight. The two cats stood for a moment looking out across the openness of the grouse moor. Within the forest they felt safe but the moorland was an alien, dangerous place, so they turned and followed the edge of the trees.

Soon the sound of a twig snapping made Samson freeze. At the edge of the forest a stag was grazing in the darkness. Delilah flattened herself into the forest floor and watched as Samson stalked forward. The stag was huge, his antlers branched into many formidable points. But the same primal instinct that told the lynx to hunt also told them that their prey was relaxed and oblivious – more asleep than awake, and unaware of the silent death that approached.

Samson moved closer, his paws barely kissing the ground. Close to the browsing stag he stopped, ears twitching, heart pounding. Some instinct raised the stag's head and he sniffed at the air and glanced around, suddenly rigid and alert; but he saw nothing, just the dark forest and the moorland in the moonlight. The stag relaxed and went back to grazing. Samson was inching closer, trembling, his whole being intent on the animal in front of him. Now he was motionless, eyes fixed on the stag. The old creature turned his head and the lynx sprang.

For a moment Samson's body hung in the air. His leap took only a fraction of a second, yet in that moment he jumped across 1,300 years and lynx were hunting in the forests of Britain once more. The impact flashed through the stag in a wave of terror. He bucked and thrashed his head back and forth, but instinct told Samson to cling fast to the writhing body, biting down hard, gripping blood-slick flesh with his claws, as he felt the antlers sweeping through the air close to him. The stag snorted, bellowed and staggered a few paces forward, eyes rolling in fear. Then he leapt high into the air and came crashing down so hard the impact almost shattered his front legs.

The shock of landing threw Samson from the animal's back and he spun sideways into the heather where he crouched for a moment, dazed. In that moment the stag, drenched in blood, turned and lunged with his antlers. Samson was inexperienced and waited too long to escape the attack. The sharp points of the stag's antlers dug into his stomach and he howled in pain. Then he was flying through the air again, but this was no controlled leap; he tumbled over and over, thrown almost into the trees by the power of the blow. Samson hit the ground with a thud once more and the air was driven out of him. Free of the cat, the stag turned and fled with astonishing speed across the moor.

Delilah padded out from the forest to where Samson lay motionless. The stag's breath still hung in the air around Samson's body. There was blood on his rear thigh from a gouge where the antlers had bitten into his flesh. But after a moment he rose, panting from exhaustion and wide-eyed from the visceral thrill of the failed hunt, and hobbled into the wood.

They covered some distance together – Samson leading, Delilah following – but with every few steps he grew weaker until, at last, he lay down, his eyes closed, and he gave himself to sleep. Delilah stood over her fallen mate not knowing what had become of him. When she licked at the tear in his thigh, Samson groaned and shivered and closed his eyes, the grass beneath him wet with blood.

⁕⁕⁕

The stocky police constable rummaged in his pockets. He knew that he had a Thornton's toffee secreted somewhere about his person but so far, in the multitude of pockets his uniform possessed, it had eluded him. He knew that if he didn't find it soon it would begin to melt and then he would come across it months later when it had become a small furry creature long past the edible stage.

Inspector Redding, with the determination of a bloodhound, was bending over and inspecting a large hole in the wire fence of the lynx enclosure.

Matthew Jameson – a tall, grey, bearded man in his fifties with an athletic bearing who wore a red fleece with the words

'Cairngorm Wildlife Park' in gold letters across the back – was shaking his head, and pointed towards the enclosure.

'They cut the lock here, you see. Then I think they must have had some sort of trailer.'

Inspector Redding nodded, caught sight of Constable Mac-Leod exploring his trouser pockets, and shot him a withering look. The constable removed his hands from his pockets. His quest for the confectionery would have to be abandoned for a little while.

With a sigh, Mr Jameson added, 'Probably lured them in with some sort of bait. Then vanished. Easy really.'

The inspector looked up from her notebook. 'Why?'

He looked pained. 'Worst-case scenario it's the fur trade. There's a lucrative black market in fur.' Mr Jameson thrust his hands deep in his pockets as he thought. 'It's unlikely to be a private collection. You can't keep animals like the lynx secret.'

The inspector pushed her hair back and carried on writing. 'No. So why else might they be stolen?'

'Who knows? Animal rights group, perhaps. Cranks.'

The police inspector picked up the remains of the lock and turned it over, examining the marks. 'Do you think these animals could survive in the wild?'

The park manager nodded. 'Oh yes, probably. We were hoping to release them anyway if the government gave permission. They weren't bred in captivity. Their mother was killed in Spain on the roads and we've kept human contact to a minimum, so I think they might survive.'

The inspector nodded and was beginning to walk away when the park manager called to her. 'One more thing.'

'Yes?'

'We think they hit a tree. The criminals, I mean. Look over here.'

The inspector bent down and examined the gash in the bark of the trunk. 'Hmm,' she said, and when the constable twigged that she was concentrating he slipped his fingers into his back pocket for the toffee.

'Green paint,' the inspector announced.

She scraped some fragments into a plastic envelope just as Constable MacLeod's fingers located a small foil-wrapped sweet in the folds of his pocket. The inspector was staring at the gravel drive now and the constable took the opportunity to slip the toffee into his large paw, undetected.

'Ah, now look at this!'

The inspector held up a small sliver of glass. As she did so Constable MacLeod secretly unwrapped the toffee, feeling its yielding softness gently pressing against his palm.

The inspector turned to the constable and held the piece of glass up in triumph. 'This is from a car headlight, I'm certain. If we can find where this glass came from we've got them.'

With that she dropped the glass into a plastic bag and smiled with glee.

Constable MacLeod watched his superior intently. Just for a second she focused her entire attention on the plastic bag and the glass. In that same moment the large constable popped the toffee into his mouth. Now he too smiled with glee, but for a very different reason.

***

An awkward silence hung in the air of the ornate wood-panelled drawing room at Castle Purdey. Gerald McCormack sat fiddling with the buttons of his waistcoat while Charles Purdey stared at him. Tabatha sat opposite, toying with her gin and tonic. She sensed that something interesting was about to happen and made herself comfortable on the chaise longue in order to watch. To say that the atmosphere between the two men was thick would be an understatement. No knife could have cut this air; a chainsaw would have been required.

Purdey nodded slowly. 'Explanation, dear chap? Yes, I rather think I would like one.'

The button McCormack was fiddling with broke off in his long, spindly fingers, and he examined it for a few seconds before depositing it in his pocket.

'It's a little embarrassing,' the councillor said quietly.

Tabatha lit a cigarette. She knew she needed one, but didn't want to miss anything when the lawyer finally spilled the beans, so she got the act done as quickly as possible. McCormack shifted in his seat and the platoon of dachshunds sitting beneath Tabatha's seat growled at the sudden movement.

'Quiet!' Tabatha yelled with such venom that the disorderly canines shrank back in fear. Even McCormack jolted in his chair with surprise.

McCormack cleared his throat and whispered, 'I'm afraid I was put under pressure.'

Purdey leaned forward. 'Little deaf – I'm sorry, what was that?'

'He said he was *put under pressure*,' Tabatha boomed in a voice that would have been heard on the summit of Ben Bhuidhe.

Purdey's eyes widened. 'Pressure? How so?'

McCormack winced, and when he spoke this time his voice was barely audible. 'A certain individual threatened to reveal something.'

Purdey shot a questioning glance at Tabatha.

'*Reveal* something, Charles,' she yelled again, rattling the window panes.

The light of comprehension shone in Purdey's eyes. 'Ah, blackmail!'

'Quite so.'

'Reveal what, exactly?'

McCormack shuffled his feet, the rubber soles of his brogues squeaking on the floor. The dogs growled again and Tabatha quietened them with a look.

McCormack glanced at Tabatha. 'I wonder if we might speak alone, Charles.'

Many of Tabatha's days passed in boredom. Now that something vaguely interesting was happening she was adamant that she wasn't going to miss it. Quickly she said, 'Charles and I have no secrets. Anything you can say to him you can say to me. Isn't that right, Charles?'

'Yes, that's quite right,' Purdey said, astutely giving the only answer that could save him from a lifetime of matrimonial retribution.

McCormack began hesitantly. 'Well, you see I ... '

Purdey cut in, no doubt to save the councillor from further embarrassment. 'Ah, I see! Don't say another word.'

'Do you, Charles?' asked Tabatha.

'Of course. There's nothing to be ashamed of these days.'

McCormack looked like a man saved from the death

sentence. 'Isn't there?'

Tabatha stubbed out her cigarette in frustration. 'Isn't there?' she echoed, glaring at her husband.

Purdey looked surprised. 'No, not at all. I'm a bit surprised at you, Gerald, that's all. Wouldn't have thought it.'

McCormack held up his hands as though surrendering. 'I am flesh and blood, you know. I suppose we all stray sometimes.'

Purdey smiled. 'I can usually spot them, but, well—'

'Charles,' Tabatha interrupted her husband, hoping to shut him up.

But Purdey was in full flow now. 'No, I would never have thought it. But it's everywhere these days. Even vicars do it.'

McCormack stared at Purdey. 'Vicars? Are you sure?'

'Charles, he's not … He's been playing away from home,' Tabatha said in a voice that could not be ignored.

Purdey looked at her, bewildered. 'What?'

McCormack nodded.

Purdey still looked confused.

Tabatha downed her gin and took a deep breath in frustration. 'Having a bit on the side.'

Purdey thought for a moment and then the light of understanding reached him from a galaxy far away. 'Ah I see. So you're not … '

Now it was McCormack's turn to be puzzled. 'Not what?'

Purdey rose. 'Time for a drink, I think.'

Tabatha sighed. Living with Charles sometimes left her feeling like the only grown-up in the room.

When they were settled, the men with their whisky and Tabatha with her gin, McCormack produced a map of the

Muir estate out of his briefcase.

'I've had an idea,' he announced, and then sat back, looking rather smug. 'Hydroelectricity.'

Purdey raised an eyebrow. 'Really?'

McCormack pointed to a place on the river. 'I got one of the council engineers to look at it. All you have to do is dam the river here. Put a road in and sit back while the money comes in.'

'But the initial investment would be considerable,' Purdey said.

'Grants, subsidies. It's green energy – the public will pay for it.'

Tabatha peered at the map. 'What about planning permission? Wouldn't there be environmental objections?'

McCormack rubbed his hands together and beamed at her. 'My dear, there are *always* environmental objections, but you can put a hydroelectric scheme anywhere you like. Look at Glen Etive. And besides, I'm chairman of the planning committee.'

He and Purdey laughed, and Tabatha joined in.

***

Delilah lay watching at the edge of the forest. Samson's body was motionless and silent for most of the next day. Only when a raven landed close by and began its stiff-legged walk towards him, looking for a meal, did he let out a low growl and scare the bird off. Once, Samson raised his head and tried to rise, but the effort was too much and he lay back in the heather just as snow began to fall from the grey sky.

At last, just as the day was giving up its light and darkness encroached from the edges of the woodland, Delilah rose and slipped away into the forest with hunger gnawing at her. She had only gone a short distance when movement caught her eye. A nervous roe deer had stepped into a small clearing, dipping its head to catch a few mouthfuls of grass and then bobbing back up to look for danger. Delilah had seen what the strength of the stag could do. Such large prey was beyond her, but this little forest deer looked like an easier and less dangerous meal.

She padded forward, checking the wind direction with the tufts of hair on her ears; the doe could not smell her. Samson had power, but in the bright gold of Delilah's eyes burned the intelligence of a hunter, and that was all she needed to equal his strength. One sound, one movement at the wrong time, and the little deer would take flight and dart through the forest with a speed she could not hope to match, leaving her staring into the vastness of the woods with hunger bitter in her belly once more.

Two more steps forward – almost close enough, but not quite. She could see the deer trembling now, feel the blood coursing through its veins, almost taste its flesh. Perhaps by some deep-buried instinct, forgotten for a thousand years, it felt the cat close by. The deer froze and looked up. Its ears swivelled, searching for the smallest of sounds, yet it did not turn and run. In her short life Delilah had never killed. She had played with her brothers and sisters in captivity, practising the moves of a hunter, and pounced on the odd mouse or bird unwary enough to enter her enclosure. This moment was different from anything she had known before, yet her instincts were keen. She tasted the wind and could feel the tiny heart of her prey beating close by. The moment had come – she tensed and leapt with all of her strength. This was not a game. Her life depended on this moment.

Her body snaked through the air, coming in low over a fallen tree. Time slowed. She watched as the thin-legged deer turned towards her, its eyes widening in alarm, its legs dipping and tensing. She watched as the fawn's body began to twist away to flee – a drama played out in the forests of these hills for 10,000 years – but all the while she hurtled closer to that beating heart. The roe deer was quick, her body and senses tuned over millennia to escape the teeth of the lynx and wolf, but not quick enough. Delilah's impact knocked her sideways and then her hooves flipped from beneath her. Once the deer was down the outcome was certain.

The creature let out one short bark of terror before Delilah's jaws found its throat and crushed its windpipe. The lynx killed, as all big cats do, by asphyxiating its prey. In less than a minute the roe deer lay dead at Delilah's feet and she began to feed.

Something stirred at the edge of the wood. Delilah looked up from her kill and growled; Samson was silhouetted against the cold light of the moorland, moving stiffly and in pain towards her. He had smelt the blood and hunger had roused him. Delilah bared her teeth and hissed when he drew closer, not willing to share the kill.

Samson made an awkward lunge, hampered by his injured leg, and lashed out at the female with his talons. Both cats faced each other, spitting and snarling. Normally Delilah would not have been able to fend off the bigger, heavier cat, but now, after the stag had hurled him into the air, he was weak and stiff and she could hold her ground. Lynx hunt alone; sharing did not come naturally to Delilah, but after a few moments of facing down the male she sheathed her claws, backed off a little, and let him feed. That night they shared the flesh of the deer and slept, bellies full for the first time, in the arms of the forest.

***

On a cold and grey morning, Donald watched with Hamish from the shelter of the heather as a stag tiptoed away from the forest, tasting the morning air as it moved. The great antlered head turned anxiously from side to side, ears and nose twitching. The beast did not feed. It spun again, head lowered, as if protecting itself from an attack from the rear. The sharp points of its antlers sliced the air but touched nothing. It trotted to higher ground, muscles quivering and tense. Donald watched the animal through his binoculars. He could see its nose twitching and its eyes rolling in panic; the stag was afraid of something, but he couldn't tell what. *What have youse seen?*

'Something's bothering him. He's nae settled,' Donald whispered to Hamish, who lay beside him in the heather.

Hamish wriggled his bulk through the heather to get closer to the older man, then flicked away a crawling tick from his sleeve. 'It can't be us. We're downwind.'

Donald bit his lip, puzzled. 'Well, there's something nae right wi' him.'

The stag was still alert and checking in every direction for threats. Donald scanned the ground around the deer but could see nothing that would cause alarm.

'What dae youse think?' Donald hissed.

Hamish shrugged. 'Might be the best chance we get. There's good ground to fire into here.'

Hamish pushed his hair back and rearranged his tweed cap. Then he unzipped the long rifle bag lying between the two ghillies and took out the gun. Carefully he set the telescopic sights of the rifle, adjusting for the range to the stag. Finally he pulled back the bolt and let a round click into the chamber.

Donald turned and waved to the German client waiting nearby, motioning him to keep down. The client stumbled through the heather. He looked as if he wasn't used to moving over rough ground.

Donald had been bringing rich men to kill stags on this ground for over twenty years now. Years ago, they had mostly come from London – city gents, bankers, lawyers, accountants and businessmen. Now it seemed that rich Englishmen were losing the urge to boast at dinner parties of how they had faced a fierce stag in the Highlands and brought it down with a .303 rifle. More and more they came from places like Germany and Belgium, even Japan. They flew in to Inverness, spent two or three days in the hills, and then returned to their air-conditioned luxury. Years ago he had been happy to do it – glad to be away from Glasgow with its teeming streets – but he took no pleasure in it now. When he first came to the glen he had seen a wild place, a place where his feet stood not on concrete but on rough heather. Now he saw a barren man-made desert used for the pleasures of wealthy men.

Donald was beginning to despise what he did, but today Johann had come all the way from Germany and paid several thousand pounds to shoot a stag, so that was what would have to happen.

Johann, a small dark-haired stockbroker from Frankfurt,

crawled twenty feet to the two gamekeepers and slipped into the gap between them. A grouse burst up from the heather as he did so, but the stag took little notice as the bird clucked and flapped its way into the air.

Hamish handed Johann the rifle. 'Just like we showed you, sir. Take your time and aim for the chest. A fine big stag he is.'

Johann was shaking, his hands sweating as he took the rifle and peered down the telescopic sight, filled with excitement and fear at what he was about to do. The German followed the line of the animal's body and found a spot where he could be fairly certain of hitting the heart. Slowly he pulled the trigger, but the gun did not fire.

Donald coughed. 'Youse might want tae take the safety catch off, sir.'

Johann nodded and flicked off the catch. The stag raised its head and turned and Donald thought for a moment that the beast had seen the three men lying in the heather. Then the shot rang out and the stag shuddered with the impact, turned and ran.

Donald was watching through binoculars. 'Aye, sir, that wis a belter, by the way.'

Johann looked puzzled, defeated by Donald's Glaswegian accent, so Hamish explained. 'He means good shot, sir.' Hamish grinned at the German, who was still shaking. 'Your first stag, sir, well done.'

Johann reached out and shook the gamekeeper's big hand, looking both elated and terrified by what he had just done, emotions confused by an instinctive bloodlust and guilt at taking a life. Donald had seen it many times before.

Donald stood, picked up the rifle, reloaded it, and clicked

on the safety catch. He walked over to where the stag had been shot and the others followed. Looking down into the heather he found some tufts of fur and drops of blood.

'He'll nae be far.'

The old keeper began following the drops of blood leading towards the forest. It didn't take them long to find the stag. The German's aim had been true and the animal lay panting at the edge of the trees, the last moments of its life ebbing away. Donald slung the rifle over his shoulder; there would be no need for a second shot.

Donald turned to Hamish. 'Ma back's stiff today. Will youse dae the gralloching?'

Hamish nodded and took a small knife from his belt. He knelt down and cut the stag's throat.

'Is it not dead?' the German asked.

Hamish lifted the hind legs of the dead deer and turned it on to its back, exposing the white belly. 'Have to cut its throat or I won't be able to get the stomach out.'

He took his knife and slit the creature's stomach open, being careful not to cut into the intestines beneath. Then he pushed his hands into the carcass and pulled out the stomach, bowels and guts of the stag, which he emptied out into the heather. The stag's entrails steamed in the cold morning air.

Hamish rose from his gory task, hands and arms bloodied to the elbow. He was about to wipe his hands when he turned to his client and smiled. 'Sorry, I nearly forgot, sir.' Then he reached out and smeared the stag's blood across Johann's forehead before standing back to admire his work. 'There you are, sir. Your first stag – you're blooded now. A small celebration, perhaps?'

But Johann had turned a light shade of green. He swayed slightly, then turned and vomited into the heather. Donald and Hamish exchanged glances. Eventually Johann righted himself and proffered a small silver hip flask with a leaping salmon on it. The two ghillies took a sip. Johann put the flask to his lips with a queasy smile, but the deep, warming aroma of the whisky must have tickled the back of his throat, for he vomited again a moment later.

Something in the forest caught Donald's eye, and he peered through the binoculars into the gloom. Just on the edge of the trees he found a dead roe. She lay partially eaten and sprawled between the trees, almost covered by fallen leaves, but Donald's eyes were keen even if his hearing had dulled.

The keeper knelt beside the deer and examined it carefully. There was something about this dead roe deer he had never seen before.

'Hamish, kin youse come and look at this?'

A moment later, Hamish was standing over the half-eaten remains of the roe deer. He shrugged. 'Fox maybe, badger?'

Donald had years more experience than the young ghillie and was carefully examining the wounds on its throat. 'Naw, I cannae see a fox daein' this. They might feed on deid things, but something brought this beast down. Something killed it.'

Hamish shrugged again. 'Dog, maybe.' His tone suggested he didn't care.

'Perhaps.'

Donald was unconvinced. He looked around the body, searching in the leaf mould for tracks. 'There! Will you look at that, now!'

But Hamish had pulled his phone out of his pocket, angling his body so that the client wouldn't see from where he stood some distance away, and now his attention was devoted to the glowing screen. With a sigh, he looked at the paw print Donald had pointed out on the ground below.

'Well?'

'Can youse nae see it, man? Ye bampot. That's a cat. A big cat!'

Hamish shrugged for a third time. 'How do you know that?'

Donald shook his head, frustrated. He couldn't believe the young ghillie could be so ignorant. 'Look at it, man. There nae claw marks. Only a cat can sheath its claws.'

These paw prints were huge. It must be a hell of a cat, Donald realised.

∗

It was already almost dark on that late autumn afternoon when the driver brought the silver minibus to a halt at the end of the gravel drive. He pressed the switch on the dashboard and the bus doors hissed open.

'See you tomorrow, Hughie,' the two girls chorused as they stepped off the bus and out into the cold night without giving him a backward glance.

'Aye,' Hughie called after them, but the two teenage girls were already engrossed in gossiping about the images on their smartphones, and if they had heard they didn't react.

Hughie smiled and slowly shook his head. It had been fifty years since he'd been the age of the two girls. So much had changed. In his school days he had lived in a hostel on the mainland, coming home at weekends on the old ferry; and the rattling school bus, which had wheezed its way up over the pass and along the twisting single-track road, had taken an hour and a half to cover the distance he had just covered in half an hour.

He'd been driving the Shiel Bus now for seven years, ferrying schoolchildren and hikers on to the island through all kinds of weather. He knew the road so well that sometimes he thought if he closed his eyes he could drive it just as well. Hughie put the bus into gear and set off for the ferry. He would be back the following morning to take the youngsters to school in Fort William. His life had settled into a rhythm, and that suited him well.

Free of its passengers, the bus powered up the steep hill out of the village. Soon he had passed the lights of Castle Purdey, a great black shape standing above its neat lawns, and was heading out into the open moorland beyond. It was dark by now and only the outlines of the hills were picked out by the full moon, already peering over the snow-covered mountaintops and shining down into the black glen below.

Hughie switched on the radio and was just in time to catch *Gardeners' Question Time*. He had no interest in gardening but he enjoyed listening to the guests visiting draughty church halls asking why their rhubarb wouldn't grow or what type of climbing rose to train up a south-facing wall. While he watched the headlights sweep out across the ribbon of tarmac that flowed into the endless blackness, Hughie fumbled in a little bag he kept in the driver's door, fingers seeking out a liquorice allsort. He felt out one of the square ones – his favourite – and popped it into his mouth.

A quiet-voiced woman, a retired schoolteacher, was worried about the health of her dahlias. Hughie was pretty sure he wouldn't know a dahlia if he fell over it and was certain he didn't care. One of the experts was about to speak.

'The answer lies in the soil,' Hughie put in, and laughed to himself.

Something moved on the road out in the darkness, a sheep maybe. Hughie braked but the animal stood fast in his headlights, unmoving. Hughie pressed harder on the brake pedal and felt the bus slowing as its tyres fought to grip the tarmac. The yellow eyes were close now, yet still the creature did not move; Hughie pressed the brake pedal down harder still, fighting to control the minibus as the vehicle bucked and swayed. Wheels squealed on the road.

For a brief moment the headlights brought the animal into sharp relief and he saw it clearly. There, standing frozen with fear, was a big cat, bigger than a domestic cat, bigger than a wild-cat, with ear tufts and eyes that glowed like pools of molten gold.

The minibus lurched and he felt the vehicle begin to spin. The headlights swept off the road and away into the dark moorland as the vehicle twisted off the road, leaned and finally turned over. Hughie watched the world spinning until the darkness came.

***

As Donald brought the Land Rover to a halt on the rough hillside track, rain battered against the windscreen.

Hamish sat beside him fiddling with his phone. 'No bloody signal again.'

Donald glanced in his mirror and saw the police four-wheel drive pull up behind him. He opened the door and got out, then fastened his tweed jacket up to his neck against the rain, pulling down the flaps of his deerstalker to cover his ears. After a moment he lit the roll-up he'd been saving behind his ear. As he sheltered behind the Land Rover, the inspector came over to him, zipping up her hi-vis police waterproof as she walked. Hamish slammed his door and joined Donald in the lee of the vehicle.

The inspector nodded shortly to Donald. Last time Donald and Hamish had encountered her, they'd been questioned about alleged wildlife crime, an incident that still rankled. Donald resolved to be polite but not friendly.

He sniffed and wiped the rain from his nose. 'Youse'll need tae walk from here.'

The inspector grimaced. 'Is it far?'

Hamish pointed to the treeline. 'No, just to those woods over there.'

The inspector turned and called to the big police constable, who looked reluctant to leave the shelter of the police vehicle. 'MacLeod, get over here.'

The constable opened the door and lumbered over to them, followed by a tall man in a red jacket with 'Cairngorm Wildlife Park' embroidered over one pocket. The rain bounced off Constable MacLeod's hat and was running in rivulets down his police-issue waterproofs by the time he arrived beside his senior officer. Donald could see that the big constable wasn't enjoying his excursion in the hills.

Constable MacLeod looked up at the sky, cast his eyes across the moor, then looked longingly back at the dry haven of the passenger seat he'd just vacated.

He turned to his inspector. 'It is a wild an' lonely place this, hey?'

The inspector scowled and turned to the man from the wildlife park. 'You all right, Mr Jameson? It's just over in those trees.'

Mr Jameson was lean and fit, but didn't look much younger than Donald. He smiled despite the deluge. 'Aye, aye, I'll be fine.'

Curtains of rain swept across the hillside as the little group plodded after Donald up towards the woods. The constable slipped and staggered in his wellies, struggling to stay upright. Donald stopped at the edge of the wood and indicated the dead deer to the police inspector.

She wiped wet hair from her face. 'Have a look, Mr Jameson, and see what you think.'

The owner of the wildlife park bent over the corpse of the deer and Donald huddled beside him. The rain spattered off Donald's deerstalker but he was oblivious to it.

'Do youse see the injuries tae the throat, sir?'

The overcast skies and the rain made it surprisingly dark in the woods. Mr Jameson produced a small but powerful torch and shone the beam on the dead animal's neck. The rain and mud had made a mess of the corpse, but the injury was clear.

'I do. It looks like the windpipe was crushed. And there's teeth marks.' Mr Jameson turned to the inspector. 'It's consistent with a big cat. Suffocating the animal by crushing the windpipe.'

Donald nodded slowly. 'That's whit I thought. I've never seen a deer died like that before. What aboot the rear? Something's been feeding on it.'

Mr Jameson shrugged. 'That's more difficult. That could have been a fox or a badger – I really couldn't say.'

The police inspector shivered in her waterproofs. 'But it could have been a lynx?'

Mr Jameson began searching the forest floor. 'If I can find a track I can say for sure.'

Donald pointed to the place where he had seen the paw print, and the wildlife park owner used his torch to examine it. It remained crisp in the soil despite the heavy rain. After a few seconds he nodded to the ghillie and waved to the police inspector.

'That's a lynx paw print. I'd stake my life on it. Probably the female, I'd say.'

***

Lord Purdey stood in his estate office, slowly turning purple. He could hardly believe what these people were telling him.

'Lynx! Lynx! Dear God, do you mean to tell me there are bloody marauding beasts running around on my estate killing everything in sight?'

Mr Jameson from the wildlife park was shaking the water from his jacket, and the police inspector stood dripping on to the office floor.

'We only know of one lynx for certain, sir,' Mr Jameson said in frustratingly reasonable tones. 'It might have been Delilah.'

'But two were stolen. Stands to reason they are both out there. My God, they'll breed!'

The irony did not escape Purdey. He had spoken out against the application to reintroduce the lynx, and the Scottish Parliament had had the sense to listen to him. That application had been turned down, and yet someone had dared to take matters into their own hands. Whoever it was had the whole of the Highlands to choose from, but they had decided to release the animals on *his* estate. Every landowner in Scotland would be laughing at him if he didn't act decisively. Purdey was certain that this was no coincidence.

The police inspector folded her notepad away and placed it carefully in her pocket, but despite her care, it was already moist and curling up at the edges. 'We don't know that, sir.'

'Right! There's only one thing to be done here,' Purdey exclaimed, stamping his feet and trembling with fury. 'Donald! Hamish!'

Donald and Hamish, who had been enjoying a few moments out of the rain drinking tea, jolted to attention.

Purdey mimed aiming a rifle and gave them a slow smile. 'This cat, boys – better get it dispatched.' He pulled the trigger on his imaginary rifle and an imaginary bullet killed an imaginary lynx a few feet away.

Hamish grinned – *good lad, that one*, Purdey thought – but Donald looked horrified and struggled for words.

'The thing is, ye see, sir, Hamish and I ... we dinnae have much experience in the field of lynx killing. Is that nae right, Hamish?'

Hamish nodded, his enthusiasm fading. 'I suppose we don't.'

'Lynx could be tricky,' Donald said. 'I'm nae sure where we'd start.'

'In the woods,' Purdey said. He could feel a vein throbbing in his forehead. 'Start in the woods, that's where you'll damn well start.'

Mr Jameson coughed politely. He didn't look intimidated or impressed. 'I think you'll find this is my lynx, and I'll not have it shot,' he said.

Briefly incandescent with rage, Purdey was about to let rip with a choice selection of expletives when a thought ambled to the forefront of his brain. He bit his tongue and smiled.

'How do you know it's your bloody lynx?'

Checkmate, zookeeper.

Mr Jameson folded his arms and glared back. 'They're not exactly thick on the ground, are they? And there hasn't been a wild one for hundreds of years. So I think we can safely say if there's a lynx out there it's one of mine.'

There followed one of those prolonged silences in which no one could decide what to say. The ticking of the antique clock on the office wall seemed to grow in volume until it filled the available air. Purdey stared down his nostrils at the inspector and the zookeeper, but they just looked right back at him.

'And the lynx is in the forest,' Donald said at length, glancing from his employer to the others. 'Well, that's Muir estate land. We cannae hunt there.'

Purdey shot the keeper a glance that would have turned a weaker man to stone. 'Muir estate land, is it? I don't think they've found those bloody deeds yet.' He beamed as if he had found the key to an ancient puzzle. 'I'll phone my solicitor. McCormack will soon solve this.'

***

An hour later, after the little gathering had dispersed, Purdey sat alone in his office. His mind was full of ways to get rid of the lynx, but the problem of the vermin not technically being on his land still vexed him. He reached into his desk drawer and pulled out a business card that belonged to the reporter, Simon Partington.

Purdey smiled. *Let's see if we can get some good publicity out of this. Man-eating cat at large, that kind of thing.*

He had his hand on the phone and was about to call the reporter when there was a knock on the office door. 'Yes.'

Hamish entered, his ginger hair in a wild mop, having dried out from the deluge on the moor. 'I wonder if I could have a word, sir?'

Purdey eyed him. He had a grudging admiration for the lad – he was useful, if a bit slow – but now was not the time. 'Yes? What can I do for you? Speak up.'

Hamish smiled, something he didn't do very often. 'What Donald was saying about the lynx, sir … '

'Yes?'

'I didn't like to say at the time, but I don't agree with him,' Hamish said quietly.

Purdey was suddenly very interested in what the young man had to say. 'Really? Why don't you sit down?'

The big ghillie sat down, and the office chair creaked under his weight. 'He's an old man, you see. I was thinking I could maybe deal with this lynx for you.'

Purdey's eyes lit up. 'Really? Shoot it, you mean?'

The ghillie looked down, as if a little ashamed of what he was about to say. 'Shooting, sir, that might draw attention.'

Purdey leaned forward, placed his elbows on the desk and steepled his fingers, keen to hear what the young man had in mind. 'Yes, I suppose it might.'

'There's quieter ways. Like traps, snares or poison. And Muir'd never know anything about it,' Hamish added quickly.

Purdey smiled. 'No one need know. Be a good bonus in it for you.'

Hamish shifted in his seat. 'Well, I was thinking ... Donald being head keeper ... He's getting on now and a bit old fashioned.'

Purdey sensed that the keeper was building up to something but couldn't decide what it might be. 'Yes?'

'Well ... if he was let go, you see,' the words came out in a rush, 'you'd be needing a new head keeper.'

Purdey understood instantly. Naked ambition was something he could relate to. He smiled. 'Of course. I'm sure you can demonstrate your worth by eliminating this cat and we could arrange a suitable promotion. As you say, Donald is getting on.'

And he is getting opinionated in his old age, Purdey thought.

Hamish nodded. He seemed relieved. 'Thank you, sir. You don't worry about a thing. I'll take care of this lynx.'

WINTER

CHAPTER 14

Light rain fell from a gunmetal sky as Rory followed Angus along the side of the river and into the forest. 'What shape did Brian say the boulder was?'

Angus turned, his beard jewelled with raindrops. 'Like a pyramid, yer ken, half in and half oot 'o the water.'

They walked on for another hour. It was hard going following the river – there was no path and the riverside was a tangle of bushes and fallen trees, the ground so soft that their feet sank into it and water puddled about their boots. The rain fell in a fine mist and though it did not appear heavy it had soaked them both through.

Rory sat down on a fallen tree. Rain seeped through his jacket and down the back of his neck. 'Let's take a few minutes.'

Angus frowned and pulled the zip up tighter on his jacket. 'I'm drookit. This rain's heavier than it looks, yer ken.'

Rory filled his water bottle from the river and they sat quietly passing the bottle between them. 'Do you think he got this far? Maybe we passed it.'

'I suppose so. We've nae passed any boulder, I'm pretty sure.' Angus took a gulp of water and passed the bottle back.

Rory peered into the forest around them. He'd never seen growth so thick. The trees leaned in to one another, in places so close that they shut out the light. The forest floor lay in perpetual darkness. The profound silence was broken only by the occasional bird call and the constant trickle of the river.

Rory finished the bottle. 'It feels like no one has ever been here. This is how this place must have been hundreds of years ago. I bet it teems with life in the summer.'

Angus turned to Rory. 'We'll have to let the police know Brian's nae dead.'

'I suppose we will. They'll have to tell his son.'

'I could tell him. I've got his number.' Angus sounded as though he thought it was his responsibility.

Rory scratched his beard. His fingers came away wet and he laughed. 'Do you think we could get our money back from the crematorium? On account of Brian not being dead.'

Angus laughed too. 'I dinnae ken if they do refunds.'

They walked on for another half an hour and were debating whether to go back when Rory noticed something big and grey in the river.

'What's that?' Rory studied the rock, looking at it from a variety of angles. From one side it had a vague resemblance to something triangular. 'Do you think this is it?'

Angus squinted at it. 'It does nae look much like a pyramid tae me.'

Rory looked at the boulder and tried to imagine how it could be the shape of a pyramid. Then he noticed a vague depression in the ground, so faint it was almost indiscernible. 'Is that a path?'

Angus followed his gaze. 'Aye, I think it is. It's an auld 'un, mind.'

The path weaved between the trees until it reached a great

mound of earth buried in the dark forest. Rory pulled back the brambles and weeds and a stone wall appeared.

'This might be it.'

Rory pulled back more undergrowth. A door emerged, covered in so much vegetation it could have been part of the forest itself. Underneath all the bushes and fallen leaves the bothy was old and decayed, like an ancient temple to a forgotten god. The forest was taking it back for itself. It had sat unvisited for perhaps thirty years and, as the trees had grown up around it, the bothy had slowly begun to sink back into the earth. The roof had almost collapsed, its ridge sinking in a great curve where the rafters had given way. Many of the tiles had fallen and lay scattered about the forest floor. Somehow the walls had remained intact, although in places they leaned backwards at odd angles. The glassless windows stared out like blind eyes.

Rory pushed against the door and it yielded with a soft creak. He fumbled for his head torch and shone it into the dark interior of the bothy. Inside it had the dank stillness of all abandoned buildings. Broken tiles and grime covered the floor, a table lay on its side with one leg missing, and the hearth – thirty years cold – had long ago spewed soot on to the floor. A faded picture of a stag somehow clung lopsided to the chimney breast. There was a sense of frozen time in an unseen world.

Rory coughed, his throat gagging with the dust he had disturbed, surprisingly thick in places despite the damp. 'Bloody hell, where do we start?'

As they searched, the years of dust rose in a thick fog and they disturbed the homes of myriad spiders. They picked through the debris of the floor, searched in the backs of warped cupboards that had been closed for decades. Rory found the remains of the bothy's old book but its pages had decayed and the names written inside had long since vanished beyond memory. After two hours they staggered outside and sat, filthy and dishevelled, blinking in the dim light of the forest.

Angus rinsed his mouth with water from his bottle, found a log to sit on, and loaded his pipe with Irish Cask. The forest was so still that the pipe smoke hung in the air before drifting gradually away into the trees.

Angus looked disconsolately at the ruin. 'That's it, then. Nothing there. At least we looked.'

'Purdey's won. Once that parasite of a solicitor gets on to it, Purdey'll take back the land. It'll only be a matter of time before his men find the lynx too, and we all know what'll happen then.' Rory sighed. 'That's the end of Glen bothy and the lynx.'

***

The police car rocked on the deck of the ferry as the *Maid of Morvan* chugged across the water towards the mainland. Rain lashed down on the roof of the car. Inside the vehicle was the distinctive fug of moist constabulary as the inspector and constable slowly dried out.

Constable MacLeod wiped the condensation off the inside of the window and wriggled a little, trying to find a dry area of trousers for his considerable posterior to sit on. The police officer consoled himself with the knowledge that at least they were off that God-forsaken hillside and on their way back to Inverness.

'It's still raining,' he remarked.

Inspector Redding gave him one of her hard looks.

'This ferry,' she said, 'is the only way on and off the island. If the lynx were brought here they had to come on this ferry.'

The constable pondered for a moment. 'What about other boats?'

She looked at him in surprise. 'Taking an interest in detective work at last, Constable? I'm impressed.' The inspector shook her head. 'I think they brought them over here on some sort of trailer. Any boat big enough to take a trailer would have to dock at the village. In a wee place like that you can't change your underpants without the whole place knowing, let alone bring wild lynx ashore. Why don't we ask the ferryman if he's seen anything? Maybe they've something on CCTV.'

Constable MacLeod could see her logic and decided that he wouldn't bother arguing with the inspector in future. She wasn't someone who liked her ideas to be questioned. He had come to realise over his four years in the force that police duties were often boring, sometimes dangerous and frequently unpleasant. As he stood in the pouring rain talking to the ferryman, he felt the water – which had been trickling down his neck for some time – finally navigate into his underwear to find a home adjacent to his scrotum. He decided that of all the days he had spent on duty, this was turning out to be one of the worst.

The ferryman peered out from beneath his hi-vis hood as a few drops of water fell from his nose. 'Aye, we have security cameras. Got to have them – it's regulations.'

The inspector managed a smile, despite the deluge. 'Then could we have a look at the footage and see who crossed on the ferry?'

The ferryman spent a moment deep in thought. 'Ah no, you couldn't do that.'

'Why not?'

'It's not worked for the last three years to my knowledge.'

Constable MacLeod realised that they were dealing with a man who hadn't the least concern about security cameras, working or otherwise.

His superior, meanwhile, drew a deep breath and continued. 'I suppose you wouldn't remember seeing two men in a Land Rover towing a trailer, would you?'

The ferryman nodded enthusiastically. 'Oh yes, I would.'

The constable opened his mouth to ask when that was, but closed it again when he saw the inspector taking out her notepad after the slightest of glances in his direction.

'And when would that be?' she asked.

The ferryman smiled. 'We get about fifty or sixty of those every day.'

Constable MacLeod tried to give the ferryman a Dirty Harry stare. If he'd been Eastwood he would have spat on the deck and walked away in disgust. He thought about spitting on the deck, but decided that the inspector was unlikely to think that cool, so simply turned away instead. Inspector Redding followed him.

MacLeod heard a laugh behind him, as if the ferryman had just remembered something, and the officers turned.

'You'll never guess what two fellas brought over behind their Land Rover last week,' the ferryman called back to them.

No, Constable MacLeod thought as he walked away, *I'll never guess because I'm past caring*.

'Growling pigs,' the ferryman yelled through the rain. 'Welsh growling pigs!'

Constable MacLeod turned, more in annoyance than curiosity. 'Welsh growling pigs? I've never seen one of those.'

'No, you won't. Nocturnal, they are. Had them covered up, he did.' The ferryman chuckled again, as if he was pleased with the joke.

Inspector Redding froze and reached for her notebook. She turned to look at the ferryman. 'So you didn't actually *see* the pigs?'

'No, I never saw them. They're nocturnal – they get upset if they see the sun – but I heard 'em growl, though.'

The inspector took out her pen. 'So when do you think you had these pigs on board?'

***

The snow was falling faster now, sweeping across the open moorland in great curtains of white. It was early for snow, but a big high-pressure system had parked itself off the north coast of Scotland and was pulling in Arctic air from the east. Donald pulled his coat tighter around him and squeezed the switch on his radio.

'Youse there, Hamish?'

There was a crackle in the silence. 'Aye. How's it going?'

Donald sniffed in the cold air. 'Nae too good. Something's spooking them. I'll walk up by the side of the woods. Maybe find one sheltering there.'

Donald shouldered his rifle and set off slowly, looking for a hind. Rich men came from all over the world and paid thousands of pounds to shoot a stag in the Highlands. They would be wined and dined in the castle, sleep in the four-poster bed, and then he and Hamish would take them out on to the hillside and find an easy target. Some took the stag's head home to have it mounted on the wall so they could spend evenings after dinner saying how they had come to Scotland to bag the beast. No one paid to kill the hinds; they had no antlers to display, and there was no glory in killing them, so that fell to Donald and Hamish. All through the cold winter months they would spend days like this out on the hills until the snow blinded them and their fingers ached with the cold.

A hind broke cover a few hundred yards away, darting nervously through the heather. The old keeper knelt and checked the wind; she could not have caught his scent. Then he pulled out his binoculars and examined the animal. She was old, past her prime. This would be her last winter. He edged closer, looking for a place to take the shot. It was difficult in the swirling snow; he needed to get closer. He found a place hidden by a small rise in the ground, shouldered his rifle, and set the sights. He placed the cross hairs over the hind's chest, just behind her front legs, and settled himself to take the shot.

The hind was oblivious to him, and then she barked once in alarm and sped away through the heather. Donald slipped the safety catch back on his rifle. 'Aw, shit.'

She couldn't have seen him, he was sure of that, yet something had frightened her. It was then that he heard it – a short mewling cry like nothing he had heard before. He didn't know what creature had made that sound, but he had been hunting all his life and knew for certain that it was an animal in pain. He rose, feeling his knee creaking – the legacy of a life on the hills.

He followed the sound towards the wood and peered into the trees through the snow. There was nothing there, but then he heard the cry again and moved closer.

The lynx saw the figure coming towards her through the snow. The wind brought his scent to her, the acrid scent of man. Instinctively she tried to run but the snare on her foreleg held her tight and pain seared through her body as she tried to pull. The figure came closer and she hissed a warning but the man only hesitated a moment before walking into the trees and stopping a few feet away.

'Dear God!'

Donald stood looking down at the cat desperately trying to escape. He could see the fear in her eyes and sense the pain of the snare; every time she pulled at her leg, he winced. Then he remembered Lord Purdey's words: 'This cat, boys – better get it dispatched.'

Delilah had given up trying to escape; now she would have to fight. She laid back her ears and snarled, her eyes gleaming bright in the forest.

Donald raised the rifle, slipped off the safety catch and took aim. He held the lynx in his sights for a long time, his finger resting on the trigger.

'Oh, tae hell with it.'

He lowered the gun and leaned it against a tree. 'Easy now.' Donald took off his heavy tweed coat and held it in front of him like a bullfighter's cape. The lynx rose up, hissed and spat at him.

Donald looked at the fangs and the claws. Even though she was not as big as a male lynx, close up and fighting for her life she was formidable.

Donald came closer and readied himself to move. 'Oh Jesus,' he muttered, fearful of the damage the claws might inflict. Then he hurled himself forward and wrapped his coat tight around the body of the big cat.

She struggled violently and almost threw him off, but Donald knew that holding on was his only chance, so he clung with all his strength. The cat trembled beneath him. Then, for reasons Donald could not understand, the cat ceased struggling. He acted quickly and, reaching forward, found the cruel wire that held the cat's paw tight. His fingers fought to loosen it but it would not budge. He knew if the cat fought again he might not have the strength left to hold her. Then the snare moved and he pulled it free. The lynx seemed to sense that he was not her enemy and her body softened.

Donald climbed to his feet and brought his heavy coat up with him. The lynx lay panting on the ground. Both of them were exhausted. They stared at each other for a moment, and then the lynx got to her feet and strolled away, soon to be swallowed by the forest. Donald picked up the snare, tore it from the ground and hurled it into the bog.

Then he felt a twinge of pain and looked down at his forearm. An evil gash had sliced almost to the bone and blood was dripping down his fingers. He hadn't felt it at the time but it must have come from the claws of the cat.

As he walked slowly out of the forest and on to the open moorland, he thumbed the radio's controls. 'Aye, youse there, Hamish?'

'Where else would I be?' came the disembodied voice of the big ghillie.

'Kin youse get yer arse up here? I've chibbed massehl on some barbed wire.'

'How the hell did you manage that?'

'Och, never ye mind. Just get up here the noo.'

Then the radio slipped from Donald's fingers. He tried to take a step but his legs buckled underneath him and he fell face first into the snowy heather.

*****

Sinead sat in her office watching the minute hand of the clock close on twelve. At precisely eleven o'clock the door swung open and Gerald McCormack entered the room. Goose pimples appeared on her skin, and it felt as though the temperature in the room had dropped a little as the lawyer entered. McCormack did not smile – what was the point? Neither did he shake the proffered hand. He merely sat down and opened his briefcase.

'Would ye like some tea?' Sinead smiled again, but her gesture was not returned.

McCormack fixed the young woman with his gaze. 'The drinking of tea, my dear, leads to more time wasted in this country than any other pursuit. I take three cups a day – one in the morning with my breakfast, one at luncheon and one before I depart to bed. Any more than that would be an indulgence.'

The room grew chillier still as the lawyer placed some old faded papers on Sinead's desk. 'These are the deeds for the Purdey estate. You will find that they are in order. Perhaps I ought to speak to Lord Muir?'

Sinead pushed the long dark hair back from her face. 'Tony isn't a lord, Mr McCormack.'

The lawyer sniffed in disapproval. Sinead wondered if he objected to the fact that he would not be dealing with the gentry, or simply because Sinead had referred to her employer by his Christian name. McCormack's frown deepened when Tony Muir arrived from working in his garden dressed in his habitual Hawaiian shirt, shorts and wellington boots, and the frown became a scowl when he had no option but to shake the landowner's compost-stained hand. Tony, Sinead was glad to see, smiled cheerfully through this performance.

The lawyer spread the deeds across the desk. 'These are the only official deeds in existence. Unless you can provide others which postdate these, then their authority is final. Do you have any deeds to the disputed area of land?'

Tony leaned forward to examine the deeds more closely. 'We can't actually find the papers, but everyone knows I won the land to the east of the river in a card game with Charles Purdey.'

McCormack smiled a thin, snake-like smile. 'I'm afraid the law is often circumspect in matters of what "everyone knows". Documentary evidence holds greater sway than rumour, and here, as you can see, is such evidence.'

The papers were yellowed with age and written longhand in time-faded ink but their import was clear.

Sinead felt fury building inside her. They had worked long and hard to get the estate to where it was and nurture the forest back to life. If Purdey got it back he would delight in felling the trees and turning everything they had created into a poisonous grouse moor. She had an overwhelming urge to punch McCormack in the throat – and if Tony hadn't been there, she probably would have. Instead, she bit her tongue.

McCormack's smile was now one of triumph. 'In one week's time Lord Purdey will assume ownership of the land to the east of the river, and I trust you'll ensure that none of your staff interferes with the land forthwith.'

'Yes, yes,' Tony said sadly and slumped into his chair a beaten man. He shivered for moment. 'Is it me, Sinead, or is it a little chilly in here?'

They put the heating on after the lawyer left, but there was a distinct chill in the room for several hours afterwards.

*****

'A little to your right, Mr Purdey,' the cameraman said, gesturing with his hand as if the aristocrat were a cat that had strayed into his garden.

Purdey smiled and stepped to his right even though he hated performing for the camera. *Got to get our point of view out*, he thought. *Can't have the nitwit public believing everything the rewilding lefties tell them.*

Tabatha, who was standing watching the proceedings, bustled over and straightened Purdey's tie. 'You look like you've been dragged through a hedge backwards.'

Purdey hated being fiddled with. 'Don't fuss, woman.'

'A bit more, sir,' said the stocky cameraman, who was dressed in combat trousers and wore a fishing waistcoat with every pocket stuffed with lenses and gear. Purdey shuffled again. 'That's perfect now. I just wanted to get that fountain of the boy … er, well, of the boy in.'

Simon, the reporter, straightened his jacket and smoothed down his hair. 'How am I?'

The cameraman peered through his lens. 'That's fine.'

Purdey adjusted his tweed jacket and beamed, feeling like an exhibit in the zoo.

The cameraman took an establishing shot of Castle Purdey and then shot a few frames of the statue of the urinating boy. 'And rolling.'

Simon walked into shot. 'Big cats aren't something you associate with the Highland countryside, but here on Morvan many locals are increasingly convinced that a pair of lynx stolen from the Cairngorm Wildlife Park have been released and are roaming the hills. Earlier I spoke to bus driver Hughie McColl, who was injured when he swerved to avoid an animal. Speaking from his hospital bed, Hughie told me he is certain that the animal he narrowly avoided was a big cat, similar to a leopard.'

The reporter walked over to Lord Purdey. 'I'm talking to Charles Purdey of the Campaign Against the Persecution of Landowners. Lord Purdey, you've seen first-hand evidence of these animals.'

Purdey put his most serious face on. 'Oh yes, indeed. Poor little roe deer was devoured by a large, ferocious predator on my estate.'

'And you think it was a lynx. Why do you think anyone would want to release these animals here?'

Purdey wrung his hands. 'It's all about this misguided notion of rewilding. There are a few people, almost none of them from the countryside, who want to bring back all manner of creatures.'

'And you think that's a bad idea?'

'The countryside just can't support animals like that. Talk to the farmers. They are the people out day after day on the land – they'll tell you it can't work. Look what happened with the sea eagle here on the west coast. They are marauding across the hills, taking lambs and even fully grown sheep. Now they are spreading east in search of prey. Scottish farming is on a knife edge, and these birds could put some sheep farmers out of business. Beavers are already flooding land. Imagine what would happen if lynx were allowed to roam free. That could be the end of Scottish farming.'

'So who do you think is doing this?'

'Well, it's vegans mostly. Very well-intentioned people, of course, but they don't understand how we do things in the country. I've been conserving wildlife like grouse and deer for over twenty years.'

Simon took the microphone back to his mouth. 'So you can shoot them?'

Purdey sighed. 'Populations have to be controlled. The dangers of lynx are already evident. Look at that school bus crash. What if it had been full of children? Action must be taken.'

'There you have it,' Simon said to the camera. 'We'll bring you the latest news on the lynx release in future bulletins.'

***

Back in his estate office, Purdey watched Tabatha feed the dachshunds before sitting down at the desk.

'So what happens now?' she asked him.

Lord Purdey slipped off his coat and sat down opposite her. He was pleased with the way his interview had gone. 'We get a licence to shoot these bloody lynx.'

'They're not bothering us,' Tabatha said as she lit a cigarette, then inhaled deeply. 'What's a few roe deer? Why are you so determined to get rid of these animals?'

'No, they are not bothering us, and that's just the point. If we allow these creatures to get established we'll have all these vegan, left-wing, bicycle-riding, rewilding cranks telling us how wild lynx have actually had no impact in the Highlands. And you know what'll happen then, don't you?'

Tabatha looked at him for a moment and blew out a long, thin stream of smoke. 'Oh God, I think I'll go to London until this is all over. I'm sick of this obsession you have with blasting things off the face of the earth.'

Purdey continued as if his wife had not spoken. 'Once they've got a toehold they'll bring back the bloody wolf, howling all over the damn place. Then it'll be protected areas everywhere and before long we'll be knee-deep in beavers. Bears, damn it!'

If Tabatha heard her husband she didn't acknowledge him. 'In fact I think I'll stay there for the rest of the winter. Soon you won't be able to take a bath in this place without terminal hypothermia or penguins watching you from the taps.'

'These people have got to understand that this is our land and they can walk on it but that's all. I won't be told what do by anyone. If I want to bulldoze tracks over my hills as far as the eye can see I'll damn well do it. *My* hills, you understand?' By now Purdey was on his feet pacing in front of the fire, purple-faced with anger. 'I'll burn every bit of heather there is because it's mine. If I want to set fire to it, I damn well will. If I and my friends want to kill every sodding mountain hare so we have enough grouse to shoot a few days of the year, we'll do it – and no one has any right to interfere.'

'I'll see you in the spring, darling. Try not to wipe out the whole of the Highlands' wildlife while I'm away, won't you, dear?'

Tabatha strode out of the room followed by her usual cloud of petulant dachshunds.

Her departure woke Lord Purdey from his reverie. 'Where are you going, dear thing?'

Tabatha's voice drifted in from the hallway, where it was rapidly receding. 'Somewhere warm, away from the sound of gunfire.'

***

Inspector Redding was tenacious. She didn't get to solve many cases that had achieved national notoriety, so she had set her sights on finding the criminals who had stolen the lynx from the wildlife park. She realised that whoever had released the lynx had to know the island very well, perhaps even intimately. Such a strong connection with the place meant that they were likely to return sooner rather than later. *Perhaps they live on Morvan*, she reflected. She thrust her hands deep into the pockets of her hi-vis jacket to keep them out of the wind. They might live on the island – that was a possibility – but the population wasn't large, only about 400, so what was the likelihood of someone local having that degree of commitment? Probably quite small. No, whoever committed this offence was unlikely to live on the island but perhaps visited it frequently. Maybe they had a holiday cottage or came for regular weekends.

If, she reasoned, they came for weekends, they might well go home on a Sunday afternoon or evening.

This logic had brought Inspector Redding and Constable MacLeod to this slipway, shivering in the wind and watching the ferry coming slowly towards them.

They were looking for a Land Rover – one very specific Land Rover, one with damage to the wing on the driver's side. It was a cold Sunday afternoon and they had waited and watched for four hours. The wind cutting down the Minch had turned the short stretch of water between the mainland and the island choppy and white. Inspector Redding kept an eye on her over-weight colleague. At first Constable MacLeod had counted the trips the small ferry took as it ploughed backwards and forwards, but at around the twelfth trip he had lost interest and now stared miserably at the shoreline in the middle distance.

He blew his nose and shuffled to keep warm. 'You'd think there'd be a snack bar.'

The inspector was watching the Shiel Bus as it waited for the ferry, a couple of glum backpackers staring out of the windows at the grey water. 'Snack bar?'

'I could murder a cup of tea and a bacon roll just now.'

A glance at Constable MacLeod's waistline indicated that the policeman had dispatched a fair number of bacon rolls in his day.

'There's always a snack bar when you get on a ferry,' he continued. 'Lochaline, ferry to Mull, snack bar there. Ardrossan, when you catch the ferry to Arran – *very* nice snack bar there.'

The inspector was beginning to tire of the constable's bacon sandwich obsession. 'Oh really? I've never been.'

By now the constable's fascination with the purveyors of fried pork to seafarers was gathering momentum. 'Orkney ferry. Gills Bay is where you get the boat. I had a sausage bap there that was delicious—'

The inspector could stand it no longer. 'All right! Why don't you just pop into that hotel up the hill and see if they'll do you a roll or something?'

The policeman was horrified. 'And leave you alone out here? What if something happened?'

The inspector cast her eyes across the slipway. A couple of works vans were loading on to the ferry, it had started to rain, and two seagulls were fighting over the discarded crust of a pork pie.

'I don't think anything has happened here for the last twenty years. I might just survive for ten minutes while you get a roll.'

'Right.'

The constable set off towards the upmarket hotel a few hundred yards from the slipway. It was the first time since they'd begun this investigation that Inspector Redding could recall him looking happy.

*＊＊

Angus handed the ticket to the ferryman through the open window of the Land Rover. The ferryman took down the hood of his waterproofs. 'How are those pigs doing, then?'

Angus was confused for a moment. 'Pigs?'

Rory leaned across to the ferryman. 'Settled in nicely, thank you.'

Now Angus was mystified. 'What?'

'Those Welsh growling pigs have settled in very nicely, haven't they, Angus?' Rory added quickly.

The ferryman grinned and nodded, and then Angus remembered the charade. 'Oh yes, the growling pigs.'

The ferryman leaned against the door of the vehicle. 'The police was asking about them. Very interested, they were. I told how they was nocturnal, and you couldn't look at them in daylight. Wrote it down, the inspector did.'

Angus felt a knot forming in the pit of his stomach, but he tried to sound as casual as he could. 'Aye? When were you talking tae them?'

By now the ferry's departure ramp was ascending behind them and they were trapped on the boat. The ferryman bustled off to collect the rest of his tickets, but he called back to them over his shoulder.

'Just this morning. You can tell them yourselves. They're watching cars come off the ferry.'

Rory gasped. 'Oh shit!'

Angus tried to be rational to quell the butterflies in his stomach. 'They could be asking questions for any number of reasons, yer ken.'

'Especially if they're interested in Welsh growling pigs,' Rory said, shrinking into his seat.

There was nowhere to run. Even if they could turn around and head back to the island, sooner or later they would have to ride on the ferry. Angus watched as the slipway drew nearer. He could see the yellow and white of the police car and a small lone figure dressed in a yellow hi-vis jacket waiting for the cars coming off. He swallowed but he couldn't get the taste of fear out of his mouth.

***

Inspector Redding watched Constable MacLeod strolling back to the slipway chewing his bacon roll. He had been over-enthusiastic with the tomato sauce, and it trickled down his chin, giving him the appearance of a constabulary vampire.

The inspector sipped her black coffee and pointed to the officer's chin. 'You might want to wipe your face.'

At that moment the Land Rover rolled slowly past them. The inspector was so engrossed in her junior officer's condiment mishap that she almost missed the green vehicle as it drove past – almost, but not quite. Her eyes caught the angular shape of the heavy old four-wheel drive as it trundled away. Redding was sharp, and she took in the two occupants in a split second: the older, stocky, grey-bearded man, and the younger, thinner man with the scraggly ginger beard. In her ten years on the force she had seen a lot of guilty people and in the instant she saw the two men she noticed them looking straight ahead as if to avoid all possible contact. That raised her suspicions – that and the fresh silver dent on the wing of the vehicle.

***

Angus stared straight ahead as he drove off the ramp. His eyes were drawn to the two police officers standing on the quayside. He was desperate not to draw attention to himself, and worried that it would show. The officers ignored Angus and Rory, and the woman seemed focused on the man's chin, which he was wiping with a piece of tissue, but then she turned and looked at them. Her eyes seemed to pierce right through Angus as they passed.

Rory turned to Angus. 'Do you think they spotted us?'

'I dinnae ken.' He kept his eyes fixed on the road.

Angus knew the answer to Rory's question, but he'd tried to deny it even to himself. They had driven around half a mile from the ferry terminal before the police car behind them switched on its blue lights. Rory and Angus exchanged anxious glances as they pulled over into the picnic area beside the loch. The inspector appeared beside the driver's door, and Angus slid back the window.

She seemed friendly enough. 'Hello, sir. Have you been far today?'

Angus tried not to let the fact that he was filled with blind panic show in his voice, but he suspected it trickled through anyway. 'We've just been to a bothy on Morvan.'

The inspector smiled again. 'Really, and which one would that be?' She had asked the question quietly, almost as if she were passing the time of day, but Angus knew there would be more difficult questions to follow.

'Glen bothy, on the Muir estate,' Angus said, trying to keep his answers as brief as possible in case the truth might inadvertently escape from his mouth.

'I see,' the inspector replied in a matter-of-fact tone. 'Do you visit the island often?'

Angus could feel the officer stalking him like a lion in the long grass. 'Aye, most weekends. For the walking and climbing.'

Rory's approach was more assertive. He craned his head over from the passenger seat. 'What's this about? We haven't done anything.'

The inspector met Rory's eyes and her attitude hardened

instantly.

'On the 13th of October, two lynx were stolen from the Cairngorm Wildlife Park. We have CCTV of two men in a Land Rover. We have reason to believe that those lynx were released on the Isle of Morvan shortly after they were taken. Can you prove where you were on the 13th of October?'

'Oh, I see,' Rory said meekly.

Angus sat with his mouth open. This police officer already knew more than she was saying.

Her tone was businesslike now. 'Would you gentlemen like to step out of the car, please?'

The pair got out. In the summer the picnic site would have been busy with children and dogs, families taking a break from long car journeys; but it was quiet now that the season was over, with just a handful of retired couples taking in the sea views from their mobile homes. Angus was glad they didn't have an audience.

The inspector brought them round to the front of the vehicle and they were joined by the large constable, who Angus thought smelt of tomato sauce. The constable stood with his arms folded, perhaps trying to look menacing.

The inspector pointed to the damage to the front wing of Angus's vehicle. 'Do you mind telling me how this happened?'

Angus and Rory spoke simultaneously; one said a wall, the other a tree.

'Swerved to avoid a deer,' Angus added. He was not a good liar, and turned red as he spoke.

'I see,' the inspector said in a way that made it clear she did see but didn't believe him.

She bent down and examined the cracked headlight minutely. Then she stood up and produced a small plastic bag with a fragment of glass inside. She looked up at Angus. 'Do you know where I found this?'

Angus didn't know but was fairly certain he wasn't going to like the answer.

'I found this,' she said, carefully extracting it from the plastic bag, 'in the driveway of the wildlife park, along with a few slivers of green paint. Let's see if it fits.'

The inspector reached out towards the headlight, her movements followed closely by the three men.

Constable MacLeod's serious façade broke and now he grinned, overcome by excitement. 'It's like Cinderella and the glass slipper.'

The inspector paused and looked at him in such a way that made it clear this was nothing like a fairy tale. She pushed the glass into the headlamp and it clicked into place, a perfect fit.

***

The drive to Inverness was long and twisting as Rory and Angus sat together in the back of the police car, silently watching the scenery pass by. After around two hours they reached the outskirts of the city and were soon pulling in to the low red-brick police station. Angus watched the automatic gates roll back as though they were entering the hideout of some 1960s Bond villain.

Inside the station they did a lot of waiting. It was eight in the evening before the young duty solicitor managed to see them. Angus thought she looked about sixteen; she was tall, with a crimson streak in her hair, and appeared bored with the whole proceedings until she read the details of the case.

She broke into a grin as she read their charge sheet, fiddling with the stud in her nose. 'Oh wow, like – theft of a pair of lynx!'

Angus nodded. He had ceased to see the funny side of it long ago.

'The thing is,' she said, 'whatever happens, don't tell them a thing. Don't answer any questions. Just say "no comment".' She gathered up her possessions and smiled. 'Two lynx, that's really cool.'

The police inspector was polite and reasonable. Angus wondered if it came as a relief to have a pair of suspects who didn't spit or swear at her.

The big constable brought them food on a plastic tray and put it down in front of them with an apologetic shrug. It had congealed long ago but chemical analysis would have revealed it had once been mashed potato, baked beans and sausages.

Rory glanced at it. 'Sorry, I'm vegan.'

The constable looked puzzled and scratched his head. 'Oh right. We can do you a cheese toastie?'

'Nae animal products,' Angus explained slowly. He could see that the constable had never heard anything like it.

The constable wrinkled his nose. 'What, no cheese? Milk?'

Rory shook his head.

'What do you have on your cornflakes?' the constable asked, astonished.

'Soya milk. Or sometimes milk made from almonds.'

'Bloody hell.' The constable thought for a long time. 'I could get you an apple.'

Rory smiled. 'That's fine.'

The constable looked pleased that he'd found something Rory could eat. He turned and walked away down the corridor, his heavy boots squeaking on the polished floor. They heard him call to the sergeant behind the desk. 'That lynx fella says they can make milk from nuts.'

They heard the sergeant laugh. 'They can do all sorts these days. They might even make you into a policeman.'

They spent that night in a cell, their shoes and belts removed in case they decided to hang themselves. The cell was empty apart from a couple of mattresses and a high window with thick armoured glass that let in light but was impossible to see through. The sound of a drunk in the next cell singing incoherently drifted through the wall.

Angus sat on the bed looking round the cell. 'I'm very sorry I got yer intae this.'

Rory was pacing across the floor like a caged animal. He stopped when Angus spoke. 'No, don't be. I'm glad we did it. Those animals would be behind bars if it weren't for us.'

'Aye, now it's us in a cage. I never thought I'd be in here, yer ken.'

Rory sat down beside Angus on the bed. 'Remember what you told me about Benny Rothman. Nothing changed until he went to prison.'

'He never stole anything like the twa of us.'

Rory shrugged. 'We didn't steal anything. We put something back, put the lynx back where they belong.'

Angus nodded; there was some truth in what Rory said. 'Aye, maybe there'll be a song aboot us.'

They both laughed but Angus didn't feel like laughing. Angus had never broken a law in his life. He never speeded, and was always careful that he was totally sober when he drove, yet here he was in a police cell waiting for court in the morning.

'I dinnae ken how Laura's going to take this.'

***

Angus and Rory sat in the dock feeling like visitors from another world, still in their hill clothes while black-gowned solicitors drifted about the court like sinister ravens. The sheriff barely looked at them while the procurator fiscal explained the charges. The young woman who had been duty solicitor the previous night in the police station was transformed, her crimson hair hidden beneath a lawyer's wig, her studs removed. Jen was sitting in one of the public benches, looking pale and worried. Angus looked around for Laura but she wasn't there. He was relieved at that; he didn't know how he was going to face her.

The sheriff was a large, broad-shouldered man who looked more like a wrestler than a judge. The young solicitor and the procurator fiscal, who represented the Crown, mumbled inaudibly to each other and the sheriff. In a few minutes it was all over.

The sheriff nodded, and the small man who was clerk of the court stood up and made an announcement. 'Bail granted, accused called to appear in one month's time.'

Moments later Angus and Rory were outside the court walking towards the old town. Jen held on to Rory's arm as if she'd never let him go. Angus couldn't remember when the air had ever felt so fresh.

The house was oddly silent when he got home. On the hall table was a small white envelope. He pushed it open. Inside was a note from Laura.

It read, quite simply: 'I don't know how to say this but I can't live with you any more.'

He walked slowly into the kitchen and put the kettle on.

CHAPTER 15

Rory could feel Jen breathing beside him on the sofa of their flat. She was thinking, he knew that. Neither of them had spoken for a while; they'd just sat and let the silence condense between them.

Jen put down her coffee cup. 'What was tha thinking?'

Rory stared at the wall. He knew he'd jeopardised their future, and sitting on the couch beside Jen that world of wild places seemed a long way away – somehow less significant.

'We just wanted to make a difference.'

Jen was angry now. 'Well tha's chuffing done that! What if ye go t'jail? What'll us do?'

Rory had no answer. He looked down at his hands.

'Tha never does owt like this. Not thee. Whose idea were this?'

'I suppose it was both of us.'

Jen got up, unable to keep still any longer. 'I can't believe Angus did summat like this. Angus! He's the one who is allus so proper, sticks to the rules.'

Rory thought back to all the conversations he'd had with Angus as they sat in front of bothy fires or walked across the hills. He remembered them watching the harrier's dance, how wild and free it had been, and then finding it dead in the heather. He remembered how angry they'd both been and how they had sat in the estate office and watched Sinead touch the dead creature with such reverence.

Rory tried to find the right words – words to help Jen understand – but it was like trying to put together the shattered pieces of an icicle. The words just seemed to melt away before he could say them.

'Angus has changed,' he said at last.

'Aye?'

Then he realised that no matter what he told her it would sound thin, somehow fake. He had to make her see.

'Do you remember when we saw that sea eagle? You know, on the way to Bearnais. How that felt. It felt kind of wonderful.'

Jen was quiet now. 'Aye, I do.'

'Well, we just wanted the lynx to be out there in the forest. Just to know they were there somewhere. Free and wild. So maybe they could have their own sky dance.'

Rory got up from the sofa and wanted so much to hold Jen at that moment. Just to feel her body against his, to let her know he was still there and that things were all right, but she tensed as he approached and he knew he could not reach out to her.

Jen picked up her coat. 'I'm on shift soon.' She turned and headed for the door. 'What about me? Ye never give me a thought, did tha? It's my life too, tha knows.'

Rory watched her pick up her keys and phone. 'It's the demo tomorrow.'

Jen looked at him, horrified. 'Ye can't go. You're on bail. If ye do owt they'll lock you up.'

Rory felt his knees go weak. He knew how much he was hurting her.

'I have to be there.'

'Aye well, please thissen. Maybe that's what I should be doing too.' Her eyes blazed with tears as she turned and walked out of the door.

Rory sat for a long time, not knowing what to do, his mind whirring with the events of the last few days. So much had happened. Now perhaps he had lost Jen. He couldn't go to work. There was no way he could get his mind to focus on the endless test scores. He didn't even know if they would want him back.

His stomach rumbled, and he realised that he had barely eaten since he'd been taken into custody. He found some crackers in their tiny kitchen and spread some of Jen's home-made hummus over them. He took a mouthful but his appetite deserted him; he threw down the plate on the coffee table and wandered over to the bookcase.

Idly, he ran his finger along the shelf, passing over the books he'd collected on the natural world and on to his small collection of mountaineering books. He stopped at one with the romantic title *The Last Hillwalker*. He'd read it before. It was an amusing tale by some old mountain man who'd written it in the years before he became a hopeless drunk living his entire life in remote Highland shelters.

Rory needed a laugh right now. He lifted the book from the shelf, but as he did so one of Jen's nursing books came with it and fell to the floor. Rory bent to pick it up and noticed an envelope sticking out.

He pulled the paper out of the envelope. It felt like high-quality paper, and he realised that it was an official letter. A solicitor's name was embossed at the top. It was addressed to Jen. 'We regret to inform you ... ' He read the letter and then saw the cheque attached. According to the date it had arrived three weeks ago. Why hadn't Jen told him? That money would change their lives. It was their ticket out of this place, just like he had been dreaming of. Why wouldn't she tell him?

Then he realised.

She doesn't want out of here at all. This is her dream.

***

Jen made her way up the stairs to their flat. She was tired after the night shift at the hospital; ten hours of changing bandages and dealing with vomit had taken their toll. The climb up the three flights of stairs seemed longer than usual. It was just after seven-thirty in the morning when she unlocked the flat door. Normally she'd tiptoe in, allowing Rory an extra half hour's sleep before he had to get up for work, but this morning she knew he wasn't home as soon as she walked in through the door. Upstairs their bed was empty; he'd got up early to get the train down to Edinburgh for the protest. He didn't need to go that early, but she knew he would have left enough time to be there for the protest even if the train was delayed or there was some other disaster. He always had to be early for everything.

Jen almost went to bed right away, but decided to go downstairs and have a cup of tea before she put on her sleeping mask and tried to get a few hours of sleep. In their small kitchen she filled the kettle and wandered into the lounge to wait for it to boil. There was a letter sitting open on the coffee table. It took her a few moments to realise it was her letter. The cheque was lying right beside it, the staple now gone.

She sat down heavily on the sofa. 'Oh shit.'

Jen had been going to tell Rory. The time had never seemed right, and then there had been the court case and everything had turned upside down. She just hadn't found the right time. Now he'd think she'd hidden it from him. *Well, I had, hadn't I? What do you call not telling him and putting it in a book?*

She heard the metallic click of the kettle switching off in the kitchen. She didn't feel like tea now and sat for a while reading the letter, wondering what Rory had felt like when he read it. She hadn't lied to him; she just hadn't told the truth. *Perhaps the truth is too much for all of us sometimes.* Now she just wanted it to go dark and to be away from everything.

As she walked to the bedroom she passed the front door and wondered if he'd ever walk through it again. She had betrayed him, denied him his dream, or even the chance of it. Maybe that was something he couldn't forgive. Perhaps she'd lost him. Did that even matter? She put on the sleep mask and found the darkness. Despite the turmoil in her head she drifted off to sleep.

***

There was a knock, but it was a long way off and she ignored it. The knock came again, closer this time, more insistent, and she realised it was someone knocking on their front door. Jen dragged herself awake and pulled the sleep mask from her face. It was half past eight.

'Just a minute,' she heard herself call, and tumbled out of bed, then pulled a dressing gown on. She was awake by the time she opened the door.

Laura looked at her standing in her dressing gown. 'Oh. I'm sorry. I never thought you'd still be in bed.'

This was Laura, but not the Laura Jen had seen before, all neat with perfect make-up. This Laura was dishevelled. She had no make-up and her eyes were puffy with tears.

Jen smiled. 'Night shift.' The nine-to-five people never understood the rhythms of nocturnal workers.

Laura looked horrified and turned to go. 'Oh, I'm sorry.'

'Nah then, get thissen in.'

Laura smiled a weak smile back, still visibly embarrassed, but followed Jen into the flat.

Jen moved through to the kitchen and put the kettle on the boil, and made the tea she'd been meaning to drink earlier. Together she and Laura sat in the lounge with steaming cups.

'He's at this protest too, is he?' Jen said to the older woman.

Laura tensed and shrugged. 'I wouldn't know. I've not been home.'

Jen was about to ask where Laura had been if she'd not been home, but decided that might be dangerous territory. 'Is tha all right?'

Laura's hands were shaking as she held her teacup. She sat in silence for a few moments as if she didn't know what to say. 'He phoned me – Willy, Angus's boss, when he didn't come into work. Wanted to know where he was. I told them I didn't know.' She giggled, nervously. 'He was very nice.'

Jen realised that BetterLife hadn't phoned, but people not turning up for work there was routine. Rory told her that people would work there for months, sometimes years, and then one day out of the blue they wouldn't show up and someone would come and quietly clear their desk. Rory had called them 'the disappeared'. She remembered laughing about it.

Now that Laura had started talking it was as if she couldn't stop. 'He's changed, you know. Has Rory changed? Angus has been going off to bothies and the hills ever since I've known him. But that's boys, isn't it? Boys have dens when they're young and they run around in woods. Most grow out of it but some never do. It's why they go to bothies.' She sipped her tea quietly and smiled at Jen. 'Big boys' dens.'

Jen smiled back, but Laura had somehow got under her skin. She felt uncomfortable with her, almost angry. 'Do you think so?'

Laura drained her tea. 'What else could it be?'

'Rory's allus been fascinated by nature. Says it's like a connection with summat real.'

She remembered watching the sea eagle with him on the hills not long after they had first met. When he'd watched that bird circling high above she had seen him come alive, seen a joy in him she'd never noticed before. She'd fallen in love with him then.

Laura sighed. 'Stealing lynx. What were they thinking of? It's in the papers, bloody idiots.'

Jen stiffened at that, felt herself become taut. 'Maybe they wanted to put something back. 'Appen that's it.'

Laura glared at Jen and swallowed. 'Rory must have put him up to it. Angus would never … '

'Maybe tha don't know what Angus would do. 'Appen none of us really knows anyone.'

Jen felt a sadness when she looked at Laura. The cosy world she lived in had been tipped upside down for reasons beyond her understanding.

Laura got up from the sofa. 'I better be off. Let you get some sleep. Poor thing.' The door closed behind her.

Jen was left alone in the silent flat. She stood for a moment and realised she was crying, hot salty tears running down her cheeks. Then she looked up into the air and could see the sea eagle high above the glen. She watched it turn, its great wings flexing as it rode the thermals, its white tail feathers spread out in the wind. Then Rory turned to her alive with joy and she knew why he had stolen the lynx – she knew what it meant to him.

Jen quickly got dressed and grabbed her phone. Perhaps there was still time. Perhaps if she was quick. She picked up her keys, stuffed the solicitor's letter into her pocket and ran out of the flat.

***

Sinead made her way down the aisle of the train, steadying herself on the arms of the seats, carrying three paper cups in a small tray. Outside, the rolling Grampian Mountains swept by. Already winter had painted the summits white and the rivers were low, the burns that fed them stilled by ice in the high mountains. The train lurched and it was all Sinead could do to save herself from hurling the contents of the cups over Angus and Rory waiting in their seats. Rory was typing on his laptop, trying to drum up more support on social media for the protest.

She put the drinks down on the sterile Formica. 'Feck me, I almost fell over on the way.' She sat down and took a sip of the brown liquid. 'Ack, that's boggin', so it is.'

Sinead glanced at Angus sitting opposite her, his newspaper spread out in front of him, his grey beard neatly trimmed, serious expression as always. Despite his gruff exterior Sinead had come to see there was a softer side to him. He had been as hurt as Sinead when the harrier had been shot. Rory was younger, more idealistic perhaps, and with his thin ginger beard and unkempt hair he could have been a poet. How strange that they could both be so moved by the plight of an animal that had died out in the wild over a thousand years before they had been born – so inspired that they were willing to risk their liberty, their careers and their relationships. They were unlikely criminals. Sinead wished that more people were like them.

Angus's gaze was fixed on the rolling grey of the Grampians as the train cut through them on its way to Edinburgh. 'I used to look at those hills and think how wild and natural they were. But I ken now what's happened to them.'

Sinead followed his gaze out on to the treeless hills, and saw the blackened areas where the moors had been burned to make the habitat better for grouse. Not far from the track a herd of deer fled from the roar of the train.

She changed the subject to lighten the mood. 'So how many folk do we think we'll get the day?'

Angus laughed. 'I dinnae ken social media. It's all passed me by. Why don't folk talk to each other?'

Rory read from his screen: 'Join us today in Edinburgh and protest outside the Scottish Parliament as they decide if the Morvan lynx are to be shot.'

Sinead smiled at Angus. 'Youse couldn't raise support like this if it weren't for the likes of Facebook.'

Rory scratched at his beard. 'The trouble is that people do talk on social media, but that's all they do. They think if they make some sort of protest on Facebook they've actually done something to change things. But all they've done is put up a post that flickers on the internet for a few seconds and is only seen by their friends.'

'Well, it's a good job they dinnae have social media in 1939, isn't it?' Angus said with a surprisingly boyish grin.

Rory mimed typing. 'Hitler has invaded Poland, sad face, sad face, crying face. Well that's sorted that then, what's for tea?'

Sinead gave up on her coffee in disgust. 'We can't all be eco-warriors like you fellas.'

Angus sighed. 'We're nae *eco-warriors*.'

'Aye? I'm thinking that if two fellas who go and steal a pair of lynx from a zoo and release them into the wild aren't eco-warriors then Robert the Bruce wouldn't have voted SNP.'

Angus was suddenly formal. 'Our legal advice is nae to go blethering aboot any such allegations until they come to trial.'

Sinead snorted. 'Youse are as guilty as a fox in a henhouse.'

'We could get time for this.' Rory was suddenly serious. 'I remember what it was like being held in that police cell overnight. I can't imagine what it would be like serving a sentence. I don't think I could do it. Just seeing steel and concrete every day. Not being able to have the wind in your face for months and months, maybe years.'

Sinead saw the distress in his face and felt sorry that she had teased him. 'Ah, it'll nae come to that. Youse'll see.'

She put her hand on Rory's shoulder. He looked up and smiled, but she thought it was a pretty weak smile.

Rory went back to his laptop. 'There's a feature in *The Scotsman*.' He turned the laptop around to show Sinead and

Angus. Then he read aloud: 'Protesters will today meet outside the Scottish Parliament to ask the first minister to use her powers to block an application by Charles Purdey for a licence to shoot lynx on his estate in the Scottish Highlands. The licence was granted by the Highland Council, but the first minister can use her executive powers to suspend it.'

Angus craned his neck to read the text. 'Lord Purdey has the support of the Campaign Against the Persecution of Landowners. A number of environmental groups are believed to be planning to march to the Scottish Parliament to ask the first minister to suspend the licence to shoot the lynx.'

***

As Rory stepped down on to the platform the sounds of train engines, loudspeakers and a thousand rushing people filled his ears. Rory never felt at ease in cities. He avoided coming to Edinburgh if possible, and had an intense desire to get out of the crowded station. He grabbed his placard with the words 'SAVE THE LYNX' painted across it and headed for the turnstiles, trying not to stab the passing people with the placard's pole.

Every August as Edinburgh hosts the Fringe performance festival its citizens take even the oddest sights in their stride. Once a year the city streets are thronged with everything from Korean martial arts dancers to the ghosts of Christmas past, so a group of protestors carrying a banner saying 'SAVE THE LYNX' didn't draw more than a passing glance.

Rory could feel the old cobbles of Edinburgh's Gothic streets pressing uncomfortably through the soles of his shoes as they walked towards the seat of Scottish government. They headed up past the tacky tourist shops, selling gaudy tartan tat to happy tourists, and on through the old town, with its stone spires and arches. It felt ironic to travel all that distance to this thronged and noisy city to save an animal that lived its life in the silent forest as far from crowds of people as it is possible to get.

Soon he could see the incongruous modernity of the parliament building, full of odd angles and shapes, looking like a Lego set assembled by a drunk. As they walked, dozens more protestors carrying banners and placards joined them, and soon the street was full. A people fired by a common objective.

A voice boomed above the crowd. 'Angus, Rory, over here!'

A man holding a placard that said 'PUT PURDEY DOWN' was jumping up and down on the pavement to attract their attention.

Angus strained to see over the crowd. 'Who's that?'

Rory, a few inches taller, managed to get a clearer view. 'I don't know, I don't recognise him.'

The man with the placard pushed his way towards them, a woman at his side. He stood before them, beaming.

'My God, it's Brian,' Rory said, unable to believe the further change in the man in front of him. It *was* Brian, but the introverted, silent man they had known was gone, replaced by a jovial-looking character.

Brian grinned again and then took both men in a big bear hug. 'You didn't think I'd let you down, did you? I'd never let you down.'

They had let Brian down. Rory remembered finding Brian's sleeping bag empty and cold that night after the party in the bothy. He remembered watching the helicopter circle the glen and feeling helpless, wondering if Brian were lying somewhere on the mountain cold and alone. Then he remembered Brian's torch finding them when they were almost spent, that night in the blizzard.

Rory hugged Brian back. 'No, Brian, I didn't think you'd let us down. I knew you wouldn't.'

The new Brian grinned back and squeezed the woman beside him. She was a large woman, perhaps in her late fifties, with a mane of curly hair and the solid look of someone who was used to doing heavy physical work outside.

'This is Kathleen.' Brian spoke as though he were introducing an angel.

Angus laughed and shook Kathleen's hand warmly. 'What hiv yer done to him?'

She smiled back. 'He just needed feeding up a bit.'

Angus stepped back and motioned the estate manager forward. 'This is Sinead, Muir's manager. Sinead, this is Brian.'

She took a careful look at him. 'Youse the fella who came back from the dead?'

'Yes, I'm sorry,' Brian said, looking a bit embarrassed.

'Don't be sorry – that's some bloody feat, so it is.'

'Brian's changed,' Rory said, aside, to Angus. 'What the hell has she been feeding him on?'

Sinead overheard and slapped him on the back. 'That's nae food, you eejit, that's love. My mammy used to say, "No pie feeds a man like a happy heart".'

The crowd pressed in around them, the noise rose, and Rory had to shout. 'Well, whatever you've been feeding him, can I have some?'

The body of people they were caught up in began to surge along the cobbled street.

Brian rummaged in his bag, holding on tight to stop himself from dropping it in the crush. 'I brought you some of my turnips.' He offered Rory the round, grapefruit-sized vegetable. 'I grew them myself. First thing I've ever grown in my life.'

Kathleen squeezed Brian's arm and looked up at him with obvious affection. 'He did well.'

Rory thought a turnip was an odd gift, but Brian seemed pleased to offer it and was so clearly proud of the first thing he'd grown. Rory was glad that this silent man had found happiness in the most unexpected of places with this reclusive woman. For the first time, Rory realised that he had been so preoccupied with the journey and plans for the protest that he hadn't been thinking about what was happening between him and Jen. Perhaps he'd pushed it from his mind. Now he looked at Brian and Kathleen and realised that they were a lot like him and Jen – or at least how they might be in thirty years' time. Brian and Kathleen had found their place on an island. He wondered if he and Jen ever would.

He remembered picking up the letter that had fallen from Jen's book. He could see it now, lying on the floor, the embossed solicitor's crest clear in his mind. He had held the cheque in his hands, turning it over and over in disbelief. It was enough money for them to find a place on an island as he had always wanted. Enough to say goodbye to BetterLife, to days of drudgery in the tiny laboratory where he worked. But Jen hadn't told him. Instead she'd hidden the cheque away, kept it secret from him. He had no secrets from her. Perhaps this was not her only secret

– were there other things she had not told him? Did she have plans for the cash that didn't include him? Maybe he had been dragging her reluctantly into his dream. He had always known that she was less committed to the idea of life on an island, and maybe he hadn't listened enough to her.

Once he began thinking about Jen and the cheque, he couldn't stop. Doubts clouded his mind.

Rory realised that Brian was still holding out the turnip for him. As he reached out to take it the crowd surged forward and Kathleen and Brian were carried off in a different direction.

They moved on until they came to the main group of protesters outside the parliament building. There were drummers beating a rhythm and the protesters were chanting, 'Our wildlife, our land. Our wildlife, our land.'

Rory couldn't tell how many folk were present, but he guessed there were at least 800 carrying banners and banging drums in support of the lynx. Eight hundred protesters – and a handful of Purdey's representatives sitting under a huge balloon in the shape of an innocent-looking lamb. Purdey was there himself, beneath his banner proclaiming the cause of the Campaign Against the Persecution of Landowners.

Men like Purdey don't usually take to the streets. They do their lobbying over expensive meals and fine wine. They do it by calling in favours, by secret threats and deals, by knowing that they can count on their kind of people to side with them when it counts. This was different, however. Purdey had his back to the wall.

A small platform with a microphone had been set up near the steps, and twenty-or-so policemen looked on. They had riot gear but their visors were up, and the officers looked more bored than concerned. Perhaps confident that the likelihood of a riot was slim, they chatted amongst themselves rather than keeping a close watch on the crowd.

Andy Warrington, the leader of Back to the Wild, the organisation campaigning for the rewilding of the lynx, was standing on the platform flanked by two Green MSPs. He coughed into the microphone.

'Thank you for coming here today to protect the lynx. For the first time in 1,300 years there are wild lynx free in the Scottish countryside.' There were cheers from the crowd. He continued. 'That these lynx were taken from captivity and released without planning or permit is something I cannot condone.'

Purdey roared from across the square. 'Bloody disgrace. Outright theft. Lock 'em up.'

Angus turned and cast a nervous glance at Rory, who felt his throat go dry.

'I cannot condone it,' Warrington said after a meaningful pause, 'but I can understand it.' Loud cheers from the crowd. 'For years, those of us who have campaigned for the rewilding of our countryside have faced ignorant opposition from generations of landowners determined to do everything in their power to lock great swathes of our land away in sporting estates for the enjoyment of them and their friends – land that should be rich with wildlife and people.'

There was another shout from Purdey: 'Quite right, hear hear!'

The crowd booed.

Lord Purdey was not having a good day. He felt powerful when he shouted loudly – especially if he was shouting loudly at people who were wrong – but there were such a lot of people out there and he was starting to feel outnumbered. Only a few of his right-minded friends had turned up. And, to make matters worse, he was just perceptive enough to realise that his enthusiastic utterances might not be working in his favour.

Donald was standing beside him on a little platform beneath the giant lamb balloon. Purdey turned to him, whispering, 'I probably shouldn't have just shouted that, should I?'

Donald looked at him then, utter contempt written in his craggy features. He had a white bandage on his arm, poking out from beneath his tweed jacket. The damned fool had managed to cut his arm on a bit of old barbed wire.

'I cannae be arsed,' Donald said slowly and clearly. 'Say what you like.'

Purdey's eyes widened. 'I'm sorry, what was that?'

Hamish leaned over to Donald and prodded him in the chest with a finger. 'Watch what you're saying, old man.'

Purdey could see naked ambition in the young keeper's eyes.

But Donald pushed Hamish away. 'I've kept my mouth shut lang enough.'

The crowd was chanting again: 'Our wildlife, our land!'

Warrington was warming to his task. 'The fact remains – despite how they got there, the lynx are free. There is no scientific evidence to show that they represent any threat to humans and only a marginal risk to livestock. We are here today to ask the Scottish government to issue a temporary order to protect the lynx in order that the impact of these magnificent animals can be measured.'

The crowd cheered.

'Rubbish!' Purdey yelled. 'These are wild animals. A danger to us all.'

A large portion of the crowd hissed back at him, and he bristled. *Blithering idiots*, Purdey thought.

***

Rory had been watching Lord Purdey's entourage carefully. The man looked even more enraged than usual, but Rory was more interested in the body language of the older ghillie: scowling, arms crossed, facing away from his master.

The reporter Simon Partington stood a few yards away from Purdey's giant inflatable monstrosity, his cameraman taking shots of the crowd and recording Warrington's speech. The reporter was typing on his phone when suddenly he looked up and caught Rory's eye. Face lighting up with recognition, he turned and said something to the cameraman, and then they both started shouldering through the crowd towards Rory and the others.

'That reporter's seen us,' Rory said. 'He must have recognised us from the court.'

Angus looked anxious. 'Say nothing about the case, yer ken.'

In an instant Simon had wriggled his way through the crowd and there was a microphone and a camera pointing at Angus and Rory. 'Here we are with the two gentlemen at the heart of this controversy. Can you tell me why you think lynx in our countryside are a good idea?'

Rory and Angus looked at each other in panic. It was Rory

who stammered into the microphone. 'We'd like the balance of nature to be restored. For the hills of Scotland to be green places, teeming with the kind of wildlife and people who used to be there.'

The reporter looked pleased at that. 'So you think the theft and release of the big cats was a good thing?'

Rory was uncertain of how much he should say. 'Well, I suppose, yes—'

Angus pushed him back from the microphone. 'We cannae discuss anything afor the court.'

A woman close by in the crowd had ceased paying attention to the speech-makers and was listening to the interview. She took hold of Angus's arm. 'Is it you? Did you release the lynx?'

Sinead stepped forward and tried to defuse the situation. 'Sure youse have the wrong fellas here.'

For a small woman, the protestor had an incredible lung capacity. Now she yelled at the assembled crowd. 'It's them, it's the ones that rescued the lynx!'

Brian – who had said more in the last two weeks than over the last two decades – now excelled himself with a great booming voice. 'Aye, that's them. They're bloody heroes.'

'*Heroes!*'

The cry went up from a small part of the crowd and then grew as if someone had released a wild animal into the mass of people. Now there were hands on Angus and Rory – not aggressive hands, but insistent hands pushing them towards the stage. The group of police, Rory noticed through the chaos, had begun to watch proceedings with a good deal more interest. A couple of them flipped down their visors. One thumbed the control on his radio.

'Heroes, heroes,' the chant went.

Rory had never spoken to a crowd before and neither to his knowledge had Angus. Now the horror of the public stage drew closer, as if they were approaching the gallows. Like twigs in a great sea, Rory and Angus were tossed towards the platform and pulled individually on to the rostrum, where they stood dripping with embarrassment.

Warrington put up his hands for silence. The crowd hushed. 'Now,' he said, 'I cannot condone the breaking of the law—'

There was a resounding boo from the crowd. Placards were raised and waved with renewed vigour, and the group of policemen got out their riot shields; perhaps this wasn't going to be such a quiet demonstration after all. The reporter and camera crew struggled to get as close to the stage as possible.

'But we should hear what they have to say,' Warrington concluded, shouting at the top of his voice to make himself heard.

***

Angus found himself holding a microphone in front of a sea of expectant faces without the faintest idea of what he was going to say. The crowd hushed in a couple of seconds. They waited for him, and then Angus caught sight of Purdey standing on the pedestal beneath his inflatable lamb, hands on hips, nostrils flared, as if he knew he was going to win.

Something inside Angus began to grow.

Now he had words. 'I mind, not lang ago, Rory and I watched the sky dance of a hen harrier. We watched yon bird sweep through the air, become part of the air. What a sight it was! I thocht it were the most beautiful thing I have ever seen. I dinnae believe in God, but that day on the hillside I ken I saw a kind of god. A god whose name is nature.' He paused. 'We saw the bird again a few weeks later soaring high in the sky, as if it was made of light. And then it was shot down from the sky by the orders of that man there.'

Angus pointed to Purdey.

Purdey broiled red and yelled back at Angus, words laden with aristocratic venom. 'That is a lie. You can't prove a thing. Bloody vegans, the lot of you.'

There was a murmur in the crowd and the beast that possessed it turned and snarled at the landowner.

'It's a damned lie,' Purdey snarled at them. 'Part of a vicious campaign against—'

But Angus was gathering pace, getting his energy from the crowd. 'We found that wonderful creature, broken and blasted tae the ground. I looked down at it and I wondered ... ' The words broke in his throat and he struggled to get them out. 'I wondered what it would take before we stood up and we said, enough o' this slaughter.'

'Absolutely nothing to do with me or my chaps,' Purdey yelled back.

The crowd ignored Purdey and applauded Angus. The energy lifted him. He felt as if he were growing as the people in the crowd clapped and cheered.

As Angus watched Purdey's entourage, Donald climbed on to the pedestal beside his lord.

'What are you doing?' Hamish said.

The old keeper smiled quietly. 'Maybe something I should've done a while ago.' He shouted to the crowd, 'If youse are looking for the min who shot the harrier then I'm here massehl.'

There was a shocked silence, and then cries of '*Shame!*'

Donald looked across at his employer. 'I'm nae proud of whit I done. I did it so there wid be enough grouse tae shoot. Well, maybe we killed enough things in these hills. Maybe we should let things live. I'm for saving these lynx frae the likes of ye, *Purdey*.'

Purdey glared at the old ghillie. 'Judas.'

Somehow the laird had found himself a megaphone, and now he clambered up on to a small stone plinth and screamed through it above the heads of the crowd. 'What do you people know? You're all damned Green-voting vegans. You don't live in the countryside. Irresponsible idealists.'

There was a shower of boos. Someone hurled a placard. It missed, but Purdey had to dodge to one side, and the police officers began to slowly move forwards.

Hamish was standing below Purdey. His earlier cockiness seemed to have evaporated, as if he had suddenly realised how isolated they were on their platform. 'Better come down now, sir – I think you might stir things up.'

But Purdey was oblivious, shaking with anger right down into his ancient plus fours. 'Bring back the lynx? Over my dead body!'

The environmental protestors murmured, and Rory yelled over to Purdey. 'Your hunting has destroyed our hills and left them treeless wastes, devoid of wildlife. It's time that changed.'

'Listen, you lentil-eating cat lover,' Purdey barked through the megaphone at him. It screeched discordantly. 'Men like me own Scotland. It's *our land* – what part of that don't you get? If we want to kill everything that moves, turn the whole damn place into a theme park, we'll bloody well do it.'

Someone from the group of protestors hurled a turnip. It struck Purdey a glancing blow and he crumbled slowly to the ground. *Just as the archaic class system the bastard represents must eventually fall,* Angus thought with a sudden grin.

'The Purdeys ne'er take a backward step!' the lord cried thinly from where he clung to the base of his stone plinth at ground level. A policeman was tugging at his sleeve, asking if he was all right, but Purdey batted him away with the megaphone.

Angus was suddenly inspired, and called out to the crowd. 'They kill thousands of mountain hares and we do nothing! They drain the land, poison it with chemicals, burn the moorland, an' we stand in silence. They poison golden eagles, shoot hen harriers, bulldoze tracks across the landscape and the government refuses to act.'

Boos of protest rose from the crowd. The police stood on the edges, riot shields poised, waiting to advance.

'Purdey's wrong,' Angus continued fiercely. 'This is *our* land, these are our hills and it is our wildlife they destroy. It has been stolen from us, yer ken.'

The crowd roared. Purdey grabbed the rope that anchored his giant balloon and hauled himself up from the ground back on to the plinth to get a view of the crowd. 'You haven't the faintest idea what you are talking about,' he yelled through the megaphone. 'I represent the landowners. We'll do what the hell we like with our own land. You're just a common criminal. A thief. Scum.'

Being reminded of the disgrace of the charges took all the energy out of Angus for a moment.

Then a voice in the crowd called, 'At least he's doing something.' There was another cheer.

Angus froze and stood silent and the crowd went quiet waiting for him to speak again. Part of him was back on the hillside watching the harrier's sky dance, then he was back in the police cell, and then he saw the lynx padding softly away into the forest. Purdey hollered something indignant over the heads of the crowd but Angus couldn't hear it, his head a whir of thought.

He felt Rory touch him on the shoulder. 'You all right, mate?'

The touch woke Angus from his dream and he spoke quietly to Rory. 'Aye, I'm all richt. I'm nae apologising to anyone.' It was a different Angus who picked up the microphone. 'All richt, I'm a criminal.'

There was a sudden hush and even Purdey lowered his megaphone. The police officers seemed as enthralled by the unfolding drama as everyone else.

'I'm guilty as charged. I took the lynx and set them free. And if you want tae lock me up for that, go ahead.' He spread his arms to the line of police.

Simon turned to his cameraman. 'Bloody hell, are you getting this?'

The cameraman nodded.

There were tears in his eyes when Angus spoke again. 'I got tired of waiting an' never seeing changes. I got sick o' seeing a few rich men holding Scotland in their back pockets and laughing at the rest of us. I got sick of the death on our hills, of seeing the place like a blood-soaked desert. I wanted tae live long enough tae see lynx back in our forests.'

Then the energy left him and he sagged at the microphone like a knocked-out boxer. It was then that they cheered for him, then that every throat seemed to open in support and the courtyard shook with applause.

It was then that the old grey ghillie picked up his hands and began to clap right in front of Purdey.

Purdey watched him in astonishment. He was standing on the plinth, arm wrapped around the rope leading up to the inflatable sheep that was straining against its mooring in the wind.

'Good God, man, you can't agree with that fool. How dare you!'

'Mibbe I should have dared a lang time ago.'

Purdey raised his megaphone and spoke again. 'You can't have lynx wandering about the place. The lot of them should be shot. What about my grouse?'

The crowd surged towards Purdey when he said that. After a moment's pause, Donald took out his pocket knife and hinged open the bright steel blade. He looked up at Purdey clinging to the rope and screaming at the crowd.

'Begging your pardon, sir.'

It's difficult to say what happened next. The rope anchoring Purdey's balloon to the ground may have come free all by itself, or perhaps Donald cut the rope in a final act of rebellion. One thing is certain: the giant inflatable lamb came free and slowly began to rise into the air.

'What the hell? What in blue blazes is going on?'

Purdey dropped his megaphone and clung in panic to the rope, which had become twisted and tangled around his arm. The balloon lurched and started to drift just above the heads of the astonished protestors. All present fell silent to watch the spectacle. After a moment there was a noise from the line of policemen – a noise that Angus could have sworn sounded like a hastily muffled laugh.

Then the wind caught the giant balloon and lifted Purdey by his entangled arm six feet into the air. Once the wind caught it, the sheep picked up speed and was soon bobbing up the Royal Mile, just above the level of the pedestrians, with Purdey wriggling on the rope frantically trying to disentangle himself.

***

Rory would later tell Angus, with great satisfaction, that a member of the aristocracy being towed through the Scottish capital by a giant sheep was not something one saw every day. Some people thought it was a publicity stunt for a tweed manufacturer, others thought the festival had started early, and some didn't know what to think, but all scattered out of the way as the screaming lord approached. An enterprising busker fired up his bagpipes and provided a few minutes of musical accompaniment. 'Flower of Scotland' was a particularly fine choice, Angus thought.

It wasn't the first time a landowner had cleared the people of Scotland out of his way with sheep, although the irony was lost on Purdey.

***

Rory stepped down from the podium with Angus. His legs suddenly felt wobbly; he was elated, exhausted, surprised and a little afraid at the same time. He turned to walk away and found Jen standing in front of him. He thought she looked small and tired, her blond hair unkempt.

She half-smiled up at him. 'Tha's done well. I'm reight impressed.'

'What are you doing here?' His words were short and sharp.

Jen's eyes filled with tears. 'I should have told thee about the money.'

'Why didn't you?'

Jen took hold of Rory's arm, but he pulled it back. 'I were going to. But I weren't sure, and I am now.'

'I thought we wanted the same things.' All his doubts returned as Rory looked at her.

Jen smiled, but it was a forlorn smile. 'Aye, well, we do.'

Maybe she'd come to the island now, but it wasn't what she really wanted. He knew that now. She'd always said maybe they'd go when they had the cash, but when the cash came she'd said nothing.

'What if I'd not found it? What then?'

As he said the words he felt himself slipping away from her, as if he were falling from somewhere very high and couldn't stop himself.

There were tears rolling down Jen's cheeks. 'Aye, I would ... I just needed time, like.'

He felt a wall slide between them. 'We don't want the same things, you and me. We don't.' He turned to walk away. 'You can keep the money.' The last words came in anger.

He left her standing in the square outside the parliament buildings and walked away into the crowd. A moment later, Angus followed him.

***

Angus slowly packed his pipe with tobacco and then watched as the match singed the surface of the mixture and smoke slowly began to rise. He was smoking in the house, something Laura would never allow, and the guilty pleasure made the smoke taste even more luxurious. Rory was sitting where Laura usually sat, finishing a can of beer. Sinead had curled up in an armchair on the other side of the room and her gentle snores provided an accompaniment to the TV, which they were watching with the sound off. Angus saw himself being interviewed briefly and then he appeared on the screen talking to the crowd.

Angus peered at the screen. 'Dae yer think that cagoule makes me look fat?'

Rory took another sip of beer. 'No.' He paused to burp softly. 'You are fat.'

Then there was a glimpse of Purdey heading off up the Royal Mile attached to the inflatable sheep.

'Where dae yer think he ended up?'

Rory shook his head. 'He's probably still going.' They laughed together and Angus pictured the aristocrat floating over the Forth Bridge.

The screen cut to Scotland's first minister, Nicola Sturgeon.

Angus grabbed the remote and turned on the sound. 'This'll be the decision.'

On the TV screen, Simon Partington flicked back his dark

hair. He seemed to have grown in confidence and stature over the last weeks, as if this remarkable story had given an injection of energy for his career.

'First Minister, what action does the Scottish government intend to take over these wild lynx on the Isle of Morvan?'

Sturgeon stared intently into the camera. 'We are aware of people's concerns and certainly have no plans to reintroduce the lynx into the Scottish countryside. However, we have received representation from Mr Muir, whose land these animals are believed to be occupying, that he is content to allow them to remain free. In view of that we intend to refuse the licence to shoot the lynx at this point. This will be reviewed in three months, which will give us an opportunity to see how things progress and what actions are required.'

Angus and Rory let out a whoop and a cheer so loud that Sinead raised herself, eyes bleary, from the chair. 'For feck's sake. What's the craic?'

Rory grinned at her. 'They postponed the decision for three months. It's a start, and if all goes well maybe they'll not let Purdey shoot the lynx when they review the situation.'

Angus had fallen silent. 'Aye, but when they look at it again the land won't be Muir's. It'll be Purdey's, and he'll be all for shooting them. *Especially* after what happened. He'll nae stop now.'

Rory sat down heavily on the settee. 'Oh yes, I forgot about that.'

Sinead stretched and took a cold chip from the takeaway box. 'Just a shame yousuns never found them deeds. Are youse sure the aul' fella said it were there?'

Rory nodded. 'He wasn't too coherent, but that's what we

understood. And something about a master of the glen.'

Sinead frowned. 'Who's that? Maybe he meant monarch of the glen.'

Then something clicked in Angus's mind.

He pictured the image of the proud stag he'd seen a thousand times hanging on pub walls and in hotel lobbies. He slapped his thigh. 'Aye! You're right. Maybe it wisnae *master*, it wis *monarch*. That old painting of a stag – yer'll have seen it on all the shortbread tins. Ah mind it were Landseer.'

At that moment an image of the old bothy came into his thoughts. It was as if he were standing in that shell of building, and then he saw it, hanging above the cold black hearth – a faded picture of a stag. *Monarch of the Glen*. Excitement coursed through him.

'There's one in the auld bothy, is there nae?'

Rory nodded. 'I think there is.'

'That's what the auld man telt us. An' we missed it. Maybe we ought to take another look.'

***

Rory shone his torch into the darkness of the ruin. Nothing had changed since their last visit. The dust they had shifted had long since dispersed, and the timeless quality of the place settled over it like an old mouldering blanket.

'There it is, up there on the chimney breast,' Rory said, pointing along his torch beam.

High up, above where the hearth once was, hung a tattered rectangle so covered in soot that no real image was visible.

Rory handed Angus the torch. 'I'll nip up and get it down.'

Rory felt for holds on the rough stone of the chimney. He found a crack with his left hand and a small ledge for his feet. He reached up for the painting, but it was still too far away. It was difficult to get good holds on the dusty old stone wall, but he inched a little higher until his fingers could just reach the edge of the old canvas. It felt precarious but he was only a few feet from the ground and reasoned he could jump back down if he needed to. The picture of the stag was only just visible, faded and mouldy, but the proud head and the huge antlers could still be discerned. Rory strained and touched the edge of the picture frame. Then the whole thing fell from the wall.

'Caught it,' Angus called. 'Careful now!'

Rory was plunged into darkness as Angus turned the torch to the picture frame in his hands. 'Give us some light, will you? I can't see a thing up here.'

Angus was too intent on the picture to heed Rory's requests. Rory watched him pull the warped cardboard backing away, and there, just like the old man had tried to tell them, was a carefully folded piece of cloth that might once have been a napkin.

'Oh Jesus,' Angus muttered. 'Look at that.'

'Have you got it?' Rory called down.

At that moment Rory felt the chimney breast he was clinging to move slightly. He froze. The masonry gave a tired groan and seemed to bulge out towards him. Suddenly it was crushing into his chest and he felt a great weight shoving him out into space. There was a rumble, distant at first but coming closer and closer, and then a giant stepped on his head and the darkness took him.

CHAPTER 16

There was a distant beeping. It sounded a long way off and was very faint but it intruded every time Rory tried to get back to sleep, like the persistent dripping of a tap.

'Rory,' Angus was calling him now; why couldn't he be left in peace? Angus's voice came again. 'Rory.'

Rory opened his eyes. He was in a white room with a bright light. Rory raised his head.

'What happened? Where am I?'

Angus was sitting beside him grinning. 'You're all richt. We're in Raigmore Hospital.'

Rory tried to sit up but his body had become incredibly heavy and he couldn't lift it. 'How the hell did I get here?'

Angus stood over him and put his hand on his shoulder. The pressure felt heavy and warm.

Rory tried to sit up again and failed. 'We better search the old bothy. I suppose we'll have to go there next week now.' He touched his head and it was swathed in bandages.

Angus laughed. 'It's all richt. We've done it. The bloody chimney fell on yer.'

Rory struggled to remember. 'Did it? You sure?'

'Yer missed the ride in the chopper too.'

There was a throbbing in his head. 'Did we find anything?'

Now, he realised, it was Jen sitting beside him. She took

his hand. 'Is tha all reight?'

Rory squeezed her hand. He remembered that he was supposed to be angry with her, but he couldn't remember why. All he knew was that he was glad to see her.

'Angus says a chimney fell on me.'

Jen laughed and ran her fingers through his hair. 'You're a daft bugger.'

Rory could feel his head clearing now. He could see Jen standing in Edinburgh where he'd left her. The details of the row crystallised in his mind. 'You know, if the island's not for you, well, maybe … '

Jen pressed her fingers to his lips and silenced him. 'I've summat for thee.'

She passed him an estate agent's brochure. Rory took it; there was a picture of a cottage surrounded by a large vegetable garden.

Rory stared at it in disbelief. 'Where?'

'Raasay. It's just a viewing, mind. But us can afford it.'

Rory searched her face. Was she doing this just for him? 'Are you sure?'

Jen nodded slowly. 'Aye. Remember when us saw the eagle?'

Rory nodded.

'Well, I reckon eagles are meant to fly. Tha can't keep 'em in a cage, can thee? And you was in a cage.'

Rory nodded slowly. 'I think I was.'

Jen smiled, her eyes moist with tears. 'We'll be reight there. An' I can commute to Skye, work in the hospital in Broadford, maybe.'

Two hours later he was walking beside Jen down the echoing

hospital corridor on his way home. He noticed Inspector Redding walking towards him in that clipped, efficient way of hers, her feet squeaking on the polished floor. When she spotted Rory she smiled, but he felt his stomach churn.

The inspector caught up with them, and when she addressed Rory she sounded cold and formal. 'Ah, I'm glad I've caught you. Heard you were in here. Building collapse – you ought to be careful.'

'Yes, I suppose.'

She looked at him and he felt her eyes taking in every detail of the bandage on his head and the bruises on his face. 'Not often an accused person admits his guilt on live TV, in front of 800 people *and* the Scottish Parliament.'

Rory sighed. 'No, I suppose not.'

'Open-and-shut case, really, don't you think?' Her smile was thin and menacing.

Rory nodded, feeling Jen take his arm.

Then a sparkle came into the inspector's eyes. 'Good job you'll not be going to trial then, isn't it?'

'What?'

'The Cairngorm Wildlife Park don't want to press charges. They wanted to release the creatures anyway. And the landowner's not complaining, so no point in taking action.' The inspector broke into a grin. 'Pity, in a way – my first high-profile case.' She turned and marched off down the corridor but stopped and turned back after a few paces. 'Oh, by the way … '

Rory didn't trust the inspector and was waiting for a trap.

'Well done. I'm all for rewilding.'

✳✳✳

Of all the seasons in the Highlands, winter is the most fickle. Some years it arrives with abrupt ferocity, turning the hills white and freezing the glens into a hard stillness. Other years it barely arrives at all, leaving the hills black and raking them with endless curtains of rain. This year the tips of the mountain summits above Castle Purdey were glazed early with an icy sheen.

The great machine stood incongruously before Purdey's mansion like an intruder from another world. Hamish climbed into the muscular yellow bulldozer and turned the key. The engine spluttered for a second, then fired, and the huge machine throbbed and rattled into life.

Purdey thrust his hands into the pockets of his tweed jacket to keep them out of the cold morning air. His smile was one of satisfaction, of anticipation.

'Marvellous.' He patted the steel bodywork affectionally. 'We'll show those vegans a thing or two, won't we, Hamish?'

Hamish's head appeared out of the window of the bulldozer, a yellow safety helmet crammed on top of his ginger hair. 'That we will, sir.'

Purdey strode around the bulldozer like a general inspecting his troops.

At that moment, Tabatha appeared from the Gothic arch of Castle Purdey's main door. She was drinking coffee from a chipped mug and was followed by her usual troop of aggravated dachshunds.

'What are you up to now, Charles?'

Purdey sniffed. 'I'm not up to anything.'

Tabatha gave her husband a suspicious look. 'Hmm, you've

got that "Purdeys ne'er take a backward step" look about you.'

'I have not.'

'You have. Like that time you tried to arrest that group of Japanese tourists who'd inadvertently wandered up the drive.'

Purdey frowned at his wife. She always brought that up when he'd only been acting within his rights. 'They'd no business being here.'

'*International incident*, the papers called it.' Tabatha searched for a cigarette but found her pockets empty. 'So what's this bloody contraption doing here?'

'They think they've got the better of me.'

'The Japanese?'

'No, the damned vegans.'

Tabatha sighed. 'Oh, it's them again.'

Purdey could never get his better half to grasp the importance of resisting the rise of these bike-riding do-gooders. Sometimes men like him had to act on principle.

'Twice. Once when the council voted the wrong way on the lynx application. And the second time in Edinburgh.'

No matter how hard he tried Purdey could not erase the ignominy of being dragged through the streets of the capital by a giant inflatable lamb.

Tabatha spoke to him as though talking to an errant toddler. 'So exactly what is the connection between people who don't eat animals and this thing?'

She kicked the bulldozer so hard that Hamish popped his head out of the cab to see if there had been any damage.

Purdey turned and looked up the hill with a steely gaze. 'That damned bothy.'

For the first time Tabatha looked alarmed. 'But it's Muir's bothy.'

'Not for long once McCormack gets here.'

Tabatha hissed, as if someone had stuck a pin in her and she was deflating. 'I wonder if it's too early for gin.'

A moment later Purdey spotted the lawyer's black Mercedes turning into the driveway. 'Ah, bang on time.'

Tabatha watched as the car drew to a halt beside the bulldozer and the lawyer and Highland councillor stepped on to the gravel of the driveway.

'Gerald!' Purdey's voice boomed out across the manicured lawn.

The lawyer smiled weakly. 'Charles, good to see you.'

'You have the papers? Are they in order?'

McCormack patted a battered leather satchel under his arm. 'Of course.' He spoke with such solemnity he might have been carrying the orders for an execution.

Purdey cast his eyes across the lawn, took in the impressive battlements of Castle Purdey, and allowed his eyes to wander to the distant horizon of the hills that his family had owned for over 300 years.

'It is a fine morning, is it not?'

McCormack nodded in agreement.

'The Purdeys ne'er take a backward step.' Purdey allowed the words to hang in the air for a few seconds, as though their utterance could summon the armies of his dead ancestors. 'Then serve them, Gerald, serve them.'

Purdey spoke the words as if he were ordering a cavalry charge. No armoured knights spurred their powerful steeds into action, except perhaps in his imagination. All that happened was that the solicitor climbed into his Mercedes and set off at a moderate speed for the Muir estate.

Tabatha watched as the lawyer drove away. 'No, not too early at all. A large one, I think.'

***

The old keeper picked up his binoculars and watched from the woods as the solicitor drove slowly past. Donald was no longer in Purdey's employment. Causing your employer to be dragged through the centre of Edinburgh while tethered to a giant inflatable sheep is something an employment tribunal was likely to take a dim view of. And Purdey would seek revenge, Donald was certain of that.

***

Sinead and Tony Muir had been watching the hands of the clock on their interminable progress towards eleven o'clock for what seemed an eternity. Tony Muir sat in his estate office, fingers steepled in front of his face, long grey hair tied back into a neat ponytail. For once his fingernails were clean and his trousers long. He looked more like an ageing rock star than a biscuit millionaire.

Sinead sat at her desk idly scrolling through emails and trying to hide the fact that she was paying no attention to anything on the screen, but simply waiting, as was her employer.

Tony Muir sighed. 'Do you know, since I became stinking rich the only thing I've learned about money is that it doesn't buy the really important things in life.'

CHAPTER 16

Sinead decided to tease him a little. 'Aye sure, though it's better than not being able to pay the rent.'

'Oh yes, I'm not saying people are happier in poverty. That must be awful. But after your first million, well … everything else is just window dressing.'

'Aye, youse right there, Tony.' Sinead tapped a few keys on the keyboard, pretending to work.

Muir leaned back in his chair with his hands behind his head, grinning. 'Come on, say it.'

She looked up from the computer. 'It must be awful for you – an estate in Scotland, villa in France, yacht on the Mediterranean. The worry of it all.'

Tony burst out laughing; he was rarely serious for long. 'Sorry, I'm just an old fool. Ridiculous, isn't it?' He smiled. 'I wonder if I should buy two gold Rolls-Royces this year.'

They both laughed. A moment later there was the crunch of gravel and McCormack brought his Mercedes four-wheel drive to a halt outside the estate office. Sinead glanced at the clock. It was one minute to eleven.

At precisely eleven the brass doorknob turned and McCormack stepped into the office. He stood for a second, looking down his aquiline nose at the landowner and estate manager. Sinead thought he looked even more emaciated than when she had last seen him, almost as if he hadn't eaten since.

Tony Muir rose from his seat and strode across the room, smiling, hand outstretched, but McCormack looked grave and didn't take the proffered paw, Muir would not accept the rebuttal. He took McCormack's hand in both his of his and shook it warmly.

'Sinead, put the kettle on – Mr McCormack's in need of tea.'

The lawyer was unmoved. 'Not for me.'

'Nonsense, man – you'll have a cup, and I won't hear a refusal.'

A steaming cup of tea was placed in the lawyer's hands and he regarded it as a man might if he had been handed a cup of cyanide.

Muir shivered. 'Sinead, it's a little nippy in here today. Would you turn the heating up a tad?'

McCormack peered through his horn-rimmed spectacles. 'To the matter in hand, Mr Muir: the ownership of the body of land in question. No deeds having been produced, the land reverts to its rightful owner, Charles Purdey, and so—'

Muir put up his hand. 'Might I stop you there for a moment?'

Sinead smiled and gently placed a shoe box in front of the lawyer, who looked up at her, confused. Sinead smiled at him again.

'Maybe ye should take a wee peek in there.'

The lawyer gingerly lifted the lid. 'It's a bit of old cloth.'

'Look a little closer,' she said. 'Think ye'll find it's a napkin, so it is.'

McCormack froze. Sinead could see that his nimble legal mind had grasped the significance of the artefact. She glanced over at Tony Muir, who shot her a wicked grin.

McCormack reached into the box with all the enthusiasm of a man about to grasp a rattlesnake. He turned the fabric over in his hands. 'Indeed, it does appear to be a napkin.'

'Now do ye think that could be the Purdey crest in the corner there?'

McCormack examined the embroidered crest. 'It appears to have the initials CP sewn into it, and … ' The lawyer's voice trailed off as he examined the Purdey crest of a peasant being flogged by a lord.

He stared at the old napkin and his lips moved as he read the words written upon it during the poker game all those years ago by the drunken hand of Charles Purdey. He read the words, looked up, tried to speak, looked down, and read them again. He reached for the cup of tea. His hands trembled and the teacup rattled, tea spilling into the saucer.

Sinead smiled at him, gently reassuring him. 'Perhaps something a little stronger, Mr McCormack?' She poured a large measure of whisky into his cup.

***

Purdey stood by the window of the great hall and looked out over the lawn and the driveway to the castle, puffing gently at a thick cigar as he waited for the solicitor to return. A large cloud of tobacco smoke surrounded him. He admired Winston Churchill and thought himself not dissimilar to his hero. Surely this was how Winston had felt waiting for the D-Day landings to commence, or hours before the Battle of Britain? Soon it would all be over and he could enjoy a brandy as he contemplated victory. This would be his sinking of the *Bismarck*, his invasion of Berlin. He and Winston had so much in common – the indefatigable spirit, the clarity of vision – and the only major difference, as far as Purdey could see at that moment, was that he couldn't paint.

He took another draw on his cigar and noticed the solicitor's car suddenly appear in the entrance to the driveway.

The black four-by-four was coming up the curve of the drive-way at considerable speed and weaving slightly. Then, to Purdey's amazement, it left the gravel and careered across the lawn towards the castle, carving great ruts in the grass. Purdey stared in horror as he saw that the statue of the boy urinating into the fountain lay directly in McCormack's path.

'Hamish!' Purdey cried. 'Get out there at once!'

There was a sickening crunch and McCormack was thrown forward into an expanding airbag. The naked boy on the plinth above tottered and then keeled over, still clutching his urinary appendage; seconds later he crashed through the windscreen and ended up face-to-face with McCormack.

The lawyer stared at the angelic face of the statue for a few seconds. 'M'lord, I must object. This is highly irregular.'

Hamish lifted the inebriated litigator out of the car and sat him on the grass just as Purdey arrived on the lawn to survey the devastation.

Purdey came puffing to a halt. 'What the blazes is going on?'

Hamish coughed. 'I can smell drink, sir.'

Purdey approached and sniffed. 'I can smell a distillery. Get him to his feet.'

Hamish lifted McCormack up until he stood swaying between them.

'Is it done? Did you serve the papers?' Purdey demanded.

At this point there was a loud yapping and Tabatha arrived, replete with her entourage of sausage dogs. 'What the hell was that bang? Oh … is he *drunk*?'

McCormack raised a finger. 'Point of order, m'lord.' Then he stared at the digit, apparently surprised to see it. 'You'll never guess. No, not in a thousand years.' He reached over unsteadily and pulled the napkin out of his briefcase.

'Look at this!'

Purdey turned slowly purple as he held the napkin in his hands. 'But this is just a napkin.'

By some miracle of physics McCormack was still upright. 'Actually, no, it's much more than that. That, sir, is a *legally binding* napkin. I'll go so far as to say that it is the only item of table linen I have ever seen that would stand up in a court of law.'

With that, gravity reasserted control and McCormack landed spreadeagled on the grass where he proceeded to sing 'Nelly the Elephant'. It is a well-known fact that, in extreme cases of inebriation, when all other musical renditions have fled the drunken mind, 'Nelly the Elephant' remains the one song everyone is able to sing.

Tabatha hurled a cigarette butt into the pool where the peeing boy had once stood. 'Well, that's the end of that.' With that she turned and headed back into the castle.

Purdey and Hamish walked away from the inebriated lawyer. The landowner's head was bowed, shoulders hunched, a dreadful rage seething in his gut. *What would Winston do?*

Hamish was climbing into the cab of the bulldozer when he called to his boss. 'I suppose we won't be needing the bulldozer after all, sir.'

Purdey froze and his teeth clenched around his cigar. Churchill would not have been defeated. 'We damned well *will* need it.'

Hamish froze on the steps. 'But it's not our land.'

Purdey laughed in defiance. 'I'll show those bloody vegans.

Once a bothy is demolished it's gone, isn't it? And once a lynx is shot it's dead. If I'm going to do this I'm going to do this in style. Hamish! Fetch me the Purdey sword and standard.'

Hamish paused, excitement beginning to grow in his eyes. 'See, with Donald gone now ... does that mean I'm head keeper?'

Purdey looked at the ginger ghillie in surprise. *Not the brightest lad*. 'Why yes, of course.'

Hamish grinned. 'I'll get the sword and standard then, sir.'

***

From his vantage point concealed on the edge of the woods, Donald had seen it all: the crash, the opening of the briefcase, Purdey stamping his feet and Hamish rushing off into the castle with a dreadful look of glee on his round face. Instantly, Donald realised what was happening. He dropped his binoculars and headed for the Muir estate.

***

When Donald broke the news, Tony Muir's mouth hung open in shock.

'What? He's going to *demolish* the bothy?'

Donald nodded. 'Aye, I think so.'

'But he's no right,' Tony stammered. He seemed genuinely flabbergasted by this development.

'Purdey's nae worried tae much about right and wrang.'

'We'll see about that.' He called over his shoulder to Sinead. 'Get the keys.'

Sinead pulled her mobile phone out of her bag. 'Maybe Angus and Rory have got a signal.'

***

Rory gazed out across the hills. From high on the ridge he could see out to the blue-green sea surrounding the island on all sides, and beyond that to the rolling mountains of the mainland with the snow-capped higher summits beyond.

Angus was finishing a sandwich as he sat on the rock beside him. 'Raasay is it, then?'

Rory nodded. 'Aye, we've finally bought the house.'

'I see. Nice island, Raasay.'

'Yes, and there's some land too.'

'Aye, that'll be good for vegetables, yer ken,' Angus said quietly.

'Rain quite a bit, I expect.'

Angus laughed. 'Oh aye, affa dreich place. Rain's good for vegetables.'

'You must have been to the bothy?'

'Aye, only once, mind. Taigh Thormoid Dhuibh.'

Rory laughed. 'That's easy for you to say. What's it mean?'

'Black Norman's hoose.'

Rory looked out across the sea again. 'Clear day.'

'Aye.'

The wind picked up. Rory went to zip up his cagoule and then changed his mind. 'Will you still come here?' he said after a pause.

'Aye, of course. Why would I nae?'

Rory nodded. 'Of course. There'll be other people in the club who'll come.' He gave a short laugh. 'Remember that night Brian found us?'

Angus chuckled to himself. 'Aye, we was drookit, yer ken.'

Rory turned to Angus. 'You know, when I've gone, I'll—'

From inside Rory's jacket his phone rang.

Angus groaned. 'Can yer nae turn that damned thing aff?'

Rory pulled the phone out of his rucksack. 'It's Sinead.'

He put the phone to his ear and his eyes grew wide with horror.

'What is it?'

Rory stood, his legs shaking. 'Purdey is going to demolish the bothy.'

***

They ran back down the mountain, tripping over stones and slipping on loose rocks as their boots struggled for purchase. Below them the long valley opened up, and there, not far from the bothy, they could see the yellow machine rumbling forward.

Angus was going as fast as he could. 'We'll nae make it.'

Rory was already well ahead of the older man. He entered the forest clearing near the bothy in time to see Purdey's machine emerge from the trees. Hamish was driving the bulldozer with Purdey standing beside him, sword raised, the Purdey standard fluttering in the air above.

Rory ran on through the trees, leaping over logs and fighting to stay upright as bushes caught his legs. Gasping for air, he looked back but couldn't see Angus. As Rory broke into the clearing the lumbering bulldozer was advancing slowly, like some great prehistoric beast, now only yards from the bothy.

Over the roar of the engine Rory heard Purdey shouting. 'It's that damned vegan. Charge!'

Hamish accelerated, and the bulldozer slammed its blade

into the corner of the bothy. The building shook with the impact. Slates fell from the roof and one window exploded inward.

Purdey stepped down from the throbbing machine and advanced towards Rory with his sword raised. He looked like he was enjoying himself. 'Get back, damn you, or I'll run you through.'

Rory stood his ground. 'You can't do this. You've no right. It's not your land or your bothy.'

'I am Charles Edward Purdey, fourth duke of Morvan,' he said, eyes narrowed in a perverse blend of joy and rage. 'My family has owned these hills for eight generations and I'll do as I bloody well please.' He lunged with the sword and missed.

Hamish yelled from the cab of the bulldozer. 'My lord, I don't think you should … '

Purdey didn't let him finish. 'I don't care what you think, you bloody peasant. Charge that bothy again.'

Hamish put the bulldozer in gear and slammed into the corner of the bothy once more. This time the huge steel blade did real damage – part of the corner collapsed, stonework tearing away in ragged chunks, spilling out on to the grass.

Rory stepped back. That sword point was dancing perilously close, and although Rory didn't seriously think that Purdey wanted to commit murder, accidents were likely when enraged posh people played with pointy things. It was time to take precautions.

He reached behind his head and tugged his ice axe from his rucksack. 'Come on, Purdey, you aristocratic arsehole. It's been coming to this for months.'

Purdey circled and waved the tip of the sword in the air.

He seemed more skilful all of a sudden, less full of blunder and bluster, and that worried Rory.

'My dear boy, it's been coming to this for 300 years.'

He lunged again but Rory parried the blow with his ice axe. Out of the corner of his eye he could see that Angus was running towards him but still had some ground to cover. The two exchanged blows, Purdey blocking the ice axe and Rory pushing the sword away. By now they were both blowing with the effort and Purdey – who was not in good shape – was tiring fast. He made one desperate lunge for Rory's head, but the hillwalker blocked the blow and sparks flew from the ice axe's steel pick. As Purdey tried to pull his sword back for another blow Rory hooked the blade with his axe and pulled it towards him. Purdey lost his footing and the pair collided and wrestled for a moment. Purdey sank his teeth into Rory's shoulder.

The hillwalker cried out in pain. 'You *bit* me, you bastard!'

Purdey laughed, rolled up his sleeve and offered his forearm to Rory. 'Here, take a chunk out of me. Go on! Oh sorry, you can't, can you – you don't eat meat.'

With that the aristocrat charged again and Rory was forced to sidestep to avoid the ancient sword. Rory swung the ice axe and the pick entered Purdey's jacket just above his shoulder. There was little resistance – the blade had missed his flesh – but as Rory pulled the axe back the sleeve came with it and Purdey was left with only one tweed-covered arm.

While the pair battled, Hamish was struggling with the gears of the bulldozer. At last he found reverse and the monster lurched backwards. By now one corner of the building had collapsed and the roof hung at a drunken angle. One more blow and the building would disintegrate entirely.

'Finish the bloody place, Hamish!' Purdey yelled and came at Rory in a blur of sweat and rage.

Rory saw the sword blade coming and hacked it away with his ice axe, but the blow was too much for the climbing tool and Rory heard the shaft crack with the impact.

Purdey smiled at that. He stood for a moment gathering his breath, then charged and swung his sword with all his might. This time the honed steel cut the axe in half. Suddenly defenceless, Rory staggered back, lost his footing and fell to the ground. The laird stood over him, sword raised, and Rory put up his hands in a futile gesture against the coming blow.

Angus arrived in the clearing at last and stood panting a few yards away. 'Purdey, for God's sake no. Think about this!'

Purdey turned to him, eyes blazing with the light of battle. 'Stay back, damn you, or I'll skewer him like a sausage.'

'You'll nae get away with this,' Angus yelled.

Purdey glared at Angus. 'Three hundred years ago my family cleared the people off this land. Then we cleared the trees, then the wildlife, and now … ' He paused to draw breath. 'Now I'm going to clear the hillwalkers and the vegans. Hamish, what are you waiting for? Finish the job.'

Hamish slammed the bulldozer into gear and was about to roll forward when a strange wailing sound filled the air. Purdey winced, and even Rory – whose attention was occupied at that moment by the sword point – cringed at the awful sound. Moments later Donald marched into the forest clearing playing his pipes at full volume. The old ghillie was using the bagpipes not as a musical instrument but for the purpose for which they were really designed: a weapon of war.

Purdey raised the sword. 'Hamish, flatten the bloody place.'

Then another sound filled the air, louder than the first – a relentless rattling and squeaking, as if a hundred rusty bicycles were approaching.

Angus turned and peered into the trees, which shook and trembled as something colossal approached. 'What the hell is that?'

Then Tony Muir's green tank lumbered into view, all cogs and steel tracks, and came to a halt a hundred yards away. It sat belching black smoke into the air and snorting angrily.

A hatch opened in the front and Muir's head popped up. He was wearing large earmuffs to protect his hearing and looked happy as a millionaire in a battle tank. 'I always say, you keep something long enough and one day it'll come in handy.'

Purdey grabbed Rory by the neck, hauled him upright, and held the sword to his throat. 'You can't threaten me with a loaded tank.'

Muir smiled goofily. 'Technically it's a self-propelled gun, but we won't dwell on that point.'

Purdey backed away a little, using Rory as a shield. 'Hamish, for God's sake, get it over with.'

Hamish stared at the barrel of the tank's gun. 'But sir, he's got a tank.'

'Self-propelled gun,' Muir corrected.

'It looks like a tank to me,' Hamish stammered.

Purdey drew himself up to his full height. 'He's bluffing. I bet the damn thing isn't even loaded.'

Muir's face fell.

Purdey crowed in triumph. 'You see! Cowards the lot of you. Hamish, forward.'

Muir shook his fist at Purdey in frustration. 'I wish it was damn well loaded, man.' With that he slapped the barrel with his fist and the world turned upside down.

***

There was a flash, a gout of flame and a roar like the end of time.

Angus found himself upside down against a tree. The air was full of smoke and falling twigs. He didn't move, but lay for a few moments reassuring himself that all his limbs were in the right place. Tentatively he moved his legs and found they still worked; surprisingly, there was no pain. Then he stood up and turned towards where the bothy might be. There was a wall of smoke and falling fragments, but no sign of the bothy, the tank or Purdey. Then Rory came staggering out of the smoke, stiff-legged and staring, like a hillwalking zombie. Angus saw Rory's mouth move but could hear no sound and realised that he was deaf.

They staggered towards each other through the debris and destruction.

Angus heard Rory say 'Bloody hell,' and realised that his hearing was coming back. Rory almost fell; Angus took him by the arm and steadied him. The acrid smoke filled their nostrils and left them coughing, eyes streaming. They walked on and a shape loomed through the smoke, nebulous at first, but slowly it took solid form. Angus realised that it was the bothy. Part of the roof was gone, and all the windows were broken, but the stone walls remained upright for the most part.

Then the yellow form of the bulldozer began to take shape.

The cab hung at an odd angle and steam was rising from what was left of the engine. All that remained of Purdey's flag was a broken shaft and a few shreds of fabric swinging limply in the breeze. There was a clatter and the cab door fell to the ground. A large shape began struggling and cursing from inside and then Hamish climbed down. His face was blackened; his clothes and the remains of his ginger hair were smouldering. He tore off his hi-vis jacket, threw it on the ground and stamped on it.

Hamish was struggling to form words. 'Och, I'm doing no more bothy demolition for you, Purdey.'

Then he turned and tried to walk but his legs wouldn't do as they were told and he stumbled and fell.

A figure emerged from the smoke. It was Donald, clutching a small pipe and a piece of tartan cloth – all that was left of his bagpipes. He looked at the tattered remains sadly and then hurled them away. 'Ach, jist when I were getting the hang of 'em an aw.'

Hamish groaned and Donald bent over and helped the smouldering ginger ghillie to his feet. He patted the younger man on the shoulder. 'Ah knew massehl that bampot would get ye intae trouble.'

Hamish swayed for a moment and began rummaging in his pockets. 'Wait, my phone. Oh, here it is.'

Hamish pulled a handful of twisted metal and plastic out of his pocket and peered at it for a few seconds. 'Hmm, no signal.'

'Come on, wee man. We better get youse hame.' Donald put his arm around the big keeper and together they walked away.

Rory kicked something and bent to pick it up. It was all that was left of Purdey's sword. The handle was intact but only a few inches of twisted metal remained of the blade.

'Purdey?' Angus said, a sad realisation forming in his mind.

Then Rory found a piece of smouldering tweed cloth.

Angus swallowed and looked at Rory. 'Oh God.'

At that moment there was a rattling at the bothy door, which had somehow remained upright against the blast. It creaked open and a blackened, smouldering man wearing bits of black cloth came limping out. As he drew closer they realised that it was Purdey. His moustache was gone – and most of his eyebrows – but it was the landowner for sure.

Purdey stood, swaying and smoking. 'Bloody vegans, with a tank. Who'd have thought it?'

Rory attempted to straighten Purdey's sword, but he managed only to bend the broken metal into the semblance of a cross and handed the remains to Purdey.

Purdey turned the twisted metal over in his hands and sighed sadly. 'Three hundred years.'

Then he looked at Angus and Rory through smoke-reddened eyes, and pointed his finger at them weakly. 'Lynx ... all right, you can have them. But no bears, draw the line.' His voice was thin and broken. 'Draw the line there. No bears.'

There was a screech of brakes and a Land Rover pulled up. Tabatha leapt out and strode over to where the three men were standing. She took in the blackened shape of her husband at a glance and turned to Angus. 'Has he been a nuisance?'

'Aye, a wee bit,' Angus said.

Purdey coughed. 'I've broken my sword.'

Tabatha looked at him with an expression of piteous anger, then took the piece of charred ironmongery from him and looked at it contemptuously. 'It's the twenty-first century, Charles.

You don't need a sword. We've got smartphones and reality TV.' Then she hurled it into the bushes.

Tabatha turned to Angus again and the light of recognition showed in her eyes. 'Ah, it's the naked man. How are you?'

'Fine, thank you.'

Tabatha smiled and sniffed. 'Come along, Charles.'

Angus and Rory watched as the pair climbed into the Land Rover.

'They had a tank,' Purdey said weakly.

'Of course they did,' Tabatha snapped, and they drove away.

'I could have *sworn* it wasn't loaded.'

Angus turned to see Tony Muir standing beside him. He'd taken his earmuffs off and was gazing, baffled, at the devastation.

A hatch in the tank opened up and Sinead peered out. 'Jasus, that was some fecking bang.'

Muir shook his head. 'But it wasn't loaded.'

Sinead smiled weakly. 'Ah well, youse never know when someone might happen along and pop a shell in fer you.'

Angus and Rory walked slowly over to the smouldering bothy and examined it. There were holes in the roof, the windows were gone, and the corner of the wall where the bulldozer had struck leaned out drunkenly.

Angus shook his head in despair. 'What are we going to tell the Mountain Bothies Association?'

Rory was silent for a moment. 'We'll tell them it was a group of yobs up from Glasgow having a party.'

'Aye,' Angus sighed. 'They'll swallow that for sure.'

***

Rory plodded through the snow for the last few hundred yards before the bothy. The night was still and incredibly dark; all he could see was his breath misting in the beam of his head torch. Snow crystals sparkled at his feet.

Later, Angus threw on another log, and the fire hissed back at him. 'Cold the nicht.'

Rory glanced around the restored bothy, watching the candles flickering above the old mantelpiece. 'This place doesn't change much, does it?'

Angus blew out a puff of Irish Cask and the room filled with the mellow cloud. 'No, that's what I like about it. You handed your notice in, then?'

Rory nodded. 'Aye, we're moving in January. I'll miss this place.'

'Maybe you'll get across now and again.'

'We've spent some nights here in front of this fire, you and me.'

Angus puffed slowly on his pipe. 'Aye, one or two.'

Rory finished his can of beer and stared thoughtfully into the fire. 'That's three months Samson and Delilah have been free.'

'They're oot in those woods right now somewhere.'

Rory opened another can of beer. 'Scotland has wild lynx again.'

'We achieved something.' Angus knocked out his pipe in the hearth.

Rory stood. 'I think I need a pee.'

'Me too.'

Outside, the moon was bright and the stars twinkled in the black vastness as the two men stood a respectable distance apart.

The wind stirred occasionally, bringing with it an icy chill.

Rory shivered and turned to go back into the bothy. As he did so a low howl came across the hills, distant but distinct.

'What the hell was that?'

Angus coughed. 'A dog.'

The howl came again, louder this time and clearer.

Rory turned and met Angus heading into the bothy. 'You sure that's a dog?'

'Aye, couldnae be anything else, yer ken.' Angus was chuckling to himself.

Then the first howl was answered by a second coming from the other side of the glen.

'That's a bloody wolf, that is,' Rory declared. Then he took a very careful look at Angus and slowly began to realise the only possible explanation. 'You never did? You never?'

Angus shook his head. 'Nae, dinnae be daft.'

There was another long, soulful howl.

Rory felt his spine tingle with excitement. 'You did, didn't you? You released a wolf.'

Angus snorted at the idea. 'Nae, what kinda fool would dae that?' Then he paused and there was silence between them for a moment. 'Well, only two or three.'

Both men laughed for a long time. From a distant hillside a wolf howled into the darkness of the frozen Highland night, and then all fell silent as the moon rose above the glen.

ACKNOWLEDGEMENTS

I would like to thank Alex Roddie of Pinnacle Editorial for his tireless support and editorial expertise, without which this book would never have come into existence.

Thanks are also due to the Ardtornish estate for offering me the opportunity to enjoy its solitude and find a place where I could hide to write this book.

I am also grateful to Austin J. Low, who kicked me in the arse when I got lazy.

FURTHER READING

If you would like to learn more about the issues raised in this book, you can find out more in the books and websites listed below.

Avery, Mark, *Inglorious: Conflict in the Uplands*
   (Bloomsbury, 2015)
Hetherington, David, *The Lynx and Us*
   (Scotland: The Big Picture, 2018)
Monbiot, George, *Feral: Searching for Enchantment
   on the Frontiers of Rewilding* (Penguin Books, 2013)
Raptor Persecution UK *www.raptorpersecutionscotland.
   wordpress.com*
Revive: the coalition for grouse moor reform *www.revive.scot*

ABOUT THE AUTHOR

John D. Burns is a bestselling and award-winning mountain writer who has spent over forty years exploring Britain's mountains. Originally from Merseyside, he moved to Inverness over thirty years ago to follow his passion for the hills. He is a past member of the Cairngorm Mountain Rescue Team and has walked and climbed in the American and Canadian Rockies, Kenya, the Alps and the Pyrenees.

John began writing more than fifteen years ago, and at first found an outlet for his creativity as a performance poet. He has taken one-man plays to the Edinburgh Fringe and toured them widely around theatres and mountain festivals in the UK.

It is the combination of John's love of the outdoors with his passion for writing and performance that makes him a uniquely powerful storyteller. His first two books, *The Last Hillwalker* and *Bothy Tales*, were both shortlisted for *The Great Outdoors* magazine's Outdoor Book of the Year. His third book, *Sky Dance*, is published in 2019. He continues to develop his career as a writer, blogger and outdoor storyteller while exploring the wild places he loves.

If you would like to follow more of John's journeys to bothies and wild places, and his adventures in the world of theatre, visit *www.johndburns.com* where you can read his blogs and listen to his podcasts.

ALSO BY
JOHN D. BURNS

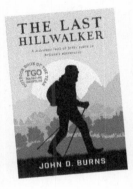

In the twenty-first century we are losing our connection with the wild, a connection that may never be regained. *The Last Hillwalker* is a personal story of falling in and out of love with the hills. More than that, it is about rediscovering a deeply felt need in all of us to connect with wild places.

In *Bothy Tales*, the follow-up to *The Last Hillwalker*, travel with the author to remote glens deep in the Scottish Highlands. Burns brings a new volume of tales – some dramatic, some moving, some hilarious – from the isolated mountain shelters called bothies.